151358

192940

Condition: F/UG+

Edition: 1st   Printing:

Jacket: w/DJ **NO**   Pages: pp(Index)

Comments: 1/2" DJ + ffe, $15

Keywords: B180

~~16.50~~ 13.50

3

FACIAL GROWTH IN THE RHESUS MONKEY

Facial Growth in the Rhesus Monkey

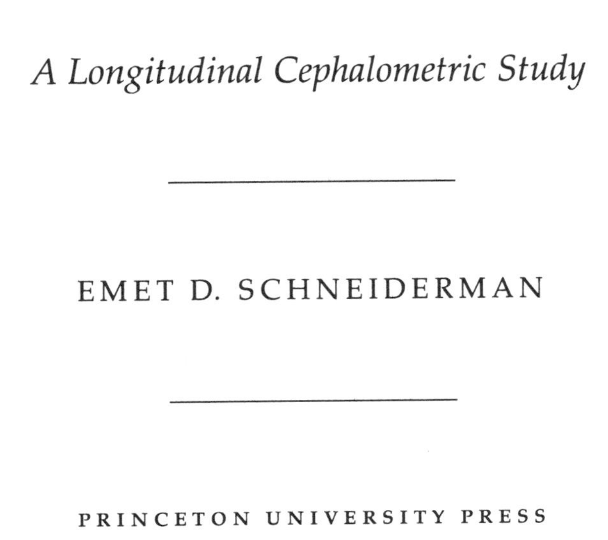

A Longitudinal Cephalometric Study

EMET D. SCHNEIDERMAN

PRINCETON UNIVERSITY PRESS

Published by Princeton University Press, 41 William Street,
Princeton, New Jersey 08540
In the United Kingdom: Princeton University Press, Oxford

Library of Congress Cataloging-in-Publication Data

Schneiderman, Emet D., 1956–
Facial growth in the rhesus monkey : a longitudinal
cephalometric study / Emet D. Schneiderman.
p. cm.
Includes bibliographical references and index.
ISBN 0-691-08749-0 (cl)
1. Rhesus monkey—Growth. 2. Face—Growth.
3. Cephalometry. 4. Longitudinal method. I. Title
[DNLM: 1. Cephalometry—methods. 2. Longitudinal Studies.
3. Macaca Mulatta—growth & development.
4. Maxillofacial Development. QL 737.P93 S359f]
QL737.P93S36 1992
599.8′2—dc20
DNLM/DLC
for Library of Congress 91-27479

This book has been composed in Linotron Palatino
Designed by Jan Lilly

Princeton University Press books are printed on
acid-free paper, and meet the guidelines for permanence
and durability of the Committee on Production
Guidelines for Book Longevity of the
Council on Library Resources

Printed in the United States of America

2 4 6 8 10 9 7 5 3 1

Dedicated to my parents

DRS. HARRIET AND LEO SCHNEIDERMAN,

whose love and encouragement have been a
guiding influence on my life.
Their example of hard work and scholarship
has given me inspiration throughout
this lengthy project.

CONTENTS

LIST OF FIGURES

LIST OF STATISTICAL TABLES (APPENDIX C)

ACKNOWLEDGMENTS

This volume would not have been possible without the significant contributions of friends, colleagues, and staff members. I thank Dr. Larry Cochard for providing one of the radiograph collections used in this study. I am grateful to the members of my doctoral dissertation committee at The University of Michigan, Drs. David S. Carlson, Charles J. Kowalski, James A. McNamara, Jr., A. Roberto Frisancho, and Milford H. Wolpoff, who have all played important roles in this work as well as in my own intellectual growth and development. Others who have made important technical contributions to this project are Joseph Mudar, Dr. Ross Anderson, Jim Stitt, Richard Miller, Vicki LaRoche, Dr. Anthony King, Carmen Banks, the late Robert Wainright, and the late Jody Ungerleider. I give very special thanks to Pearl Kapuscinski for her diligent effort and editorial assistance required to produce this manuscript. Lastly, I thank my wife, Ann McCann, for her love, encouragement, and tolerance, and my daughter, Hannah Rose, for occasionally allowing me to work on this book.

This research was supported in part by grants DE05232, DE03610, and DE08730 from the National Institutes of Health.

FACIAL GROWTH IN THE RHESUS MONKEY

1

INTRODUCTION

The rhesus monkey (*Macaca mulatta*) has been widely used for more than three decades as an experimental animal in the study of craniofacial growth. Despite its widespread use, adequate standards of normal craniofacial growth for the full range of chronological ages for this species are not available. This study seeks to alleviate this deficiency by applying statistically rigorous methods to a relatively large collection of longitudinal cephalometric radiographs of rhesus monkeys in order to provide a coherent description of the growth and maturation of the facial skeleton of this long-lived primate.

The specific purpose of this investigation is to describe dimensional and positional changes, as well as osseous remodeling, within the craniofacial complex of the rhesus monkey as they occur over the first ten years of life. To accomplish these ends, three major methodological problems had to be overcome: (1) a biologically sound approach had to be devised to quantify the morphological information contained in the serial radiographs; (2) a practical means of accurate age estimation was needed; (3) an informative and accurate longitudinal statistical model was required for representing the morphological changes as a function of time. Thus the development and application of these methods, in addition to the basic biological finding to which they lead, constitute major portions of this study.

CHAPTER 1

SIGNIFICANCE

A basic quantitative description of normal facial growth in a nonspecialized higher primate species, based on long-term longitudinal records, will be useful to a wide variety of investigators, as no comparable information produced with appropriate methods is available for any primate species (Sirianni and Swindler 1979). Good estimates of dimensional variability in the skull of any primate species *as a function of age* are currently lacking. Besides the studies by McNamara and Graber (1975) on mandibular growth and by McNamara and coworkers (1976) on maxillary growth—both studies were limited because they provided standards using four gross developmental categories—plus a brief preliminary report on craniofacial growth in pigtailed macaques (Sirianni 1985), there are no good long-term longitudinal studies in which *localized sites of growth and remodeling* have been considered in nonhuman primates. The current concepts of craniofacial growth of the rhesus monkey are based largely on cross-sectional histological studies of dry skulls (Enlow 1966; Duterloo and Enlow 1970). To be of practical use, these concepts must be reevaluated and quantified by means of a thorough longitudinal investigation.

Standards of normal craniofacial growth consisting of good estimates of central tendency and variability are needed to interpret the numerous experimental studies that use the rhesus monkey as a primate model for human growth. Baseline data are sorely needed, in particular, to help make sense of experiments concerned with the effects of orthodontic and orthognathic surgical intervention in actively growing as well as nongrowing monkeys (Smith and Minium 1983). Thus, a thorough understanding of *normal* growth is a prerequisite to the understanding of abnormal as well as therapeutically altered growth. Furthermore, approaches are needed for evaluating time-dependent phenomena (such as growth, treatment effects, and adaptation to surgery) that can be applied readily to any longitudinal data, whether from humans, monkeys, or from any other laboratory animals.

A comprehensive description should be useful to primatologists and paleontologists as well as to those investigators explicitly doing experimental studies on growth and adaptation in the rhesus monkey. This sort of information based on living primates could also aid in the interpretation of fossil primate material. For instance, it would be advantageous to have a sound quantitative basis for distinguishing the effects of variation due to age or sex differences from those due to taxonomic differences.

APPROACHES

Strategies for gaining biologically meaningful and objective information from lateral head X-ray films (cephalograms) have been elusive over the sixty-year history of radiographic cephalometry. By expanding on the implant techniques first pioneered by Björk (1955a, 1955b), we have been able to develop a satisfactory system for describing the remodeling and repositioning of the elements of the facial skeleton relative to one another and to the cranium. By using a computer-aided system, we can investigate local sites of growth as well as simple dimensional changes in the craniofacial complex. Without this in vivo bone-marking technique, an accurate longitudinal description of bone remodeling is impossible.

Aside from the problems associated with obtaining an appropriate sample with which to analyze normal longitudinal growth in the rhesus monkey, the statistical approaches for such an analysis have also been lacking, at least in application. Therefore, a major focus of this project was the development and application of appropriate statistical methods that account for intercorrelations among serial observations found within longitudinal data sets. Rao's polynomial growth curve-fitting approach (Rao 1959; Schneiderman and Kowalski 1985) and

Hills' related approach (Hills 1968; Schneiderman and Kowalski 1989) were developed as suitable alternatives to the widely used but potentially very misleading least-squares methods in the analysis of longitudinal data. The recent accessibility of high-speed computing and powerful matrix algebra programming languages (SAS 1982a, 1982b, and *GAUSS* [Edlefsen and Jones 1985]) have made it possible to implement the most rigorous statistical methods available that until now, due to the computational complexity, have been applied only to trivial examples of data in textbooks.

In the process of developing biometric and statistical methods for characterizing craniofacial growth, it was necessary to develop procedures for estimating chronological age, since this information was unknown in many of the wild-caught animals used in this study. By devising a means for calculating reliable estimates of chronological age, it was possible to generate age-specific standards for the continuous range of ages from infancy through adulthood. Such an approach was required for documenting subtle short-term developmental changes in morphology. These refinements in technique have also made possible a number of insights into sexual dimorphism.

SCOPE OF PRESENTATION OF THIS STUDY

The digitized data set collected for this investigation consists of an enormous amount of morphological information on the craniofacial complex, including the entire facial skeleton, cranial base and vault, and dentition. The scope of this monograph is limited, however, to morphological change within the maxilla and mandible and their positional relationships to each other and the cranial base.

Since development of the radiographic cephalometric technique using implants was primarily a methodological concern, the presentation of this topic is confined to chapters 1 and 2. The age-estimation problem required resolution to address the other areas considered in this investigation, yet is an issue important in its own right. Also, a separate sample of rhesus monkeys was used for this part of the project. For these reasons, the age estimation problem is presented in a separate chapter (chapter 3), with its own background, results, and discussion sections.

This study has focused on growth in males since adequate long-term cephalometric data were available only for this sex. Thus, the detailed biostatistical descriptions of maxillary and mandibular growth in chapter 4 are based exclusively on the males. The more limited female sample was described in order to examine sexual dimorphism throughout growth. The statistics based on the small female sample must be viewed with caution since they were produced, out of necessity, with less rigorous methods. Therefore, the findings and discussions regarding sexual dimorphism must be viewed as tentative.

BACKGROUND

Numerous cross-sectional investigations have been conducted on ontogenetic variation in the skulls of higher primates since the 1860s (see review in Sirianni and Swindler 1979). Despite the availability of roentgenographic cephalometry since the 1930s, this powerful technique, which permits the longitudinal investigation of morphological change, was not applied to nonhuman primates until the 1950s.

Baume (1951a, 1951b, 1951c) presented radiographic, craniometric, and histologic findings on maxillary and mandibular growth, as well as various aspects of dental eruption in forty-one rhesus monkeys. The extent to which these investigations were truly longitudinal, as well as the precise manner in which these methods were applied to arrive at numerous conclusions, is not clear from these brief abstracts, which were apparently never published as articles. These factors make it impossible to assess the reliability of the various claims.

By using radiographic cephalometry in conjunction with radiopaque implants, Gans and Sarnat (1951) were able to describe short-term longitudinal growth at the zygomaticotemporal and zygomaticomaxillary sutures, as well as appositional growth on the zygomatic arch and zygomatic process of the frontal bone in eight infant/juvenile rhesus monkeys. Moore (1949) and Craven (1956) analyzed histologically the maxillary complexes of young rhesus monkeys stained vitally with alizarin dyes. These studies, in conjunction with that of Baume (1951a), identified the maxillary tuberosity as a major region of growth responsible for the forward displacement of the maxilla.

Erickson (1958) and Pihl (1959) investigated normal mandibular growth in three and two young rhesus monkeys, respectively, using radiographic cephalometry in conjunction with radiopaque implants. Both of these studies were unpublished master's theses and are therefore unavailable. Turpin (1968) reported on mandibular growth in two young rhesus monkeys using histological sections of specimens stained with tetracycline. Two of the animals used in this study also had radiopaque gold wire implants and therefore were also studied radiographically. Sites of mandibular remodeling were analyzed in three unaltered and four partially edentulated infant rhesus monkeys using similar methods by Michejda and Weinstein (1971). Kanouse and coworkers (1969) evaluated mandibular condylar growth (specifically, mitotic activity) in four rhesus monkeys of various ages using autoradiography.

Enlow (1966) described remodeling and displacement of the bones of the facial skeleton of the young rhesus monkey. Findings were based on ground sections prepared from the dry skulls of eleven monkeys. Duterloo and Enlow (1970) described remodeling and displacement of the cranial bones in similar study using fifteen dry monkey skulls. Although these papers present a reasonable picture of the growth and development of the entire skull of the rhesus monkey, it must be recognized that they are *qualitative* descriptions and are based solely on cross-sectional material. In these studies, the "growth history" of bones was reconstructed by examining the arrangement of layers of the various types of bone tissues (Enlow 1966). Although it is an informative means of investigating bone growth, this sort of analysis alone cannot provide truly normative data, as it has not been subjected to statistical treatment and may defy it. No objective measures of central tendency and variability were reported in these studies. Additionally, there was no serious consideration of how patterns of remodeling may change throughout the long period of active growth in the rhesus monkey.

The use of radiopaque implants in conjunction with radiographic cephalometry in the present study provides a suitable approach for rigorously evaluating Enlow's (1966) and Duterloo and Enlow's (1970) descriptions of craniofacial growth. The implants provide a means of quantifying the behavior of various sites of remodeling in the skull, as well as what these investigators have designated *growth movements,* that is, translations of the various components relative to one another.

Enlow and coworkers (1971) and Enlow (1982) have popularized the concept of compensatory remodeling and growth, that is, depository and resorptive activity occurs in specific regions of the craniofacial skeleton so as to maintain certain spatial relationships and consistencies of form of the various component bones. On the basis of the pattern of correlations found within a cross-sectional human data set, Solow (1966) has promoted the related concept of the *dentoalveolar compensatory mechanism.* Although they are intuitively sensible concepts, they require quantitative corroboration that can be provided by cephalometric analysis of serial radiographs using bone implants.

An additional problem with these studies is that they did not consider sexual dimorphism; the sex of the skulls used is not even mentioned. The present study does address differences in growth between the sexes.

Elgoyen and coworkers (1972) reported on changes within the maxilla and mandible of thirteen juvenile monkeys. McNamara (1972) extended observations on normal growth in the rhesus monkey to additional maturational groups including infants, adolescents, and adults. Data were available for animals in two 13-week study periods; twenty-eight of the monkeys were followed during the first of these periods, and sixteen were followed over the second. Both of these studies used cephalometric radiography with implants for recording morphological change.

McNamara and Graber (1975) provide a report specifically on the normal mandibular growth of rhesus monkeys. This study also used cephalometric radiography with implants. McNamara and coworkers (1976) reported on normal maxillary growth in rhesus monkeys in a related study, which also used a mixed-longitudinal design. In both these studies, means and standard deviations for each of four gross dental age categories (infant, juvenile, adolescent, and young adult) were presented. The present study, which reexamines many of the same monkeys, seeks to expand on these earlier reports by providing more rigorous measures of central tendency and variability for the *continuous* range of ages from infancy to adulthood.

The continuous distribution of observations in the present study provides insights into the sources of sexual dimorphism that could not be detected in earlier studies on craniofacial growth in rhesus monkeys because of the nature of the available samples (McNamara and Graber 1975, and McNamara et al. 1976). In these earlier studies, the female sample was lacking during the period in which sexual dimorphism was greatest. In addition, the method of combining all observations for each of the four developmental categories into four sets of means tended to collapse longitudinal variability.

Carlson and coworkers (1978) provided a detailed qualitative and quantitative histological study of normal mandibular condyle cartilage of rhesus monkeys at various maturational levels. In a radiographic cephalometric study using metallic *muscle* implants in addition to bone implants, Carlson (1983) described the normal growth and migration of the masseter muscle in relation to the associated craniofacial skeleton. These findings at the condyle and in the masseter are considered as they relate to the present longitudinal cephalometric findings regarding other sites of mandibular remodeling, as well as dimensional and positional changes (particularly rotational) of the mandible that occur during maturation.

Sirianni and Van Ness (1978) have provided an excellent description of growth and development of the cranial base in the pigtailed macaque, *Macaca nemestrina*, using radiographic cephalometry in conjunction with tantalum implants placed in the basicranium. Thirty-two animals of both sexes were observed at various intervals, characterizing the growth of this region from approximately one to six years of age. These investigators noted that the cranial base angle becomes more obtuse due to anterosuperior repositioning of nasion and posterosuperior repositioning of basion. The relocation of basion was attributed to differential growth at the spheno-occipital synchondrosis. Furthermore, they concluded that the rate and duration of growth at this synchondrosis was relatively greater in males than in females between 3 and 5 years of age. The somewhat longer and flatter cranial base observed in the males was attributed to these growth differences.

In the intervening time since the present study was first published as a doctoral dissertation (Schneiderman 1985), several related studies on facial growth in the rhesus and other macaques have been published. Sirianni and Swindler (1985) published a comprehensive treatise on growth and development of the pigtailed macaque. This monograph is based on a mixed-longitudinal sample of 140 monkeys, 70 of each sex, followed from birth to as old as 9 years of age. They report on a set of sixteen standard cephalometric measurements as well as a great variety of skeletal, dental, and somatic data, using standard least-squares regression and *t*-tests for analysis.

Luder (1987a) reviewed evidence for and presented new data regarding the pubertal spurt in the growth of the mandibular condyle in nonhuman primates. This study includes histological data on eight crab-eating macaques, *Macaca fascicularis* (see also Luder 1987b), and longitudinal cephalometric data on two rhesus monkeys. These latter two monkeys were derived from the same sample used in the present study. Nielsen and colleagues have recently presented several studies (Nielsen et al. 1989a, 1989b; Bravo et al. 1989) that expand upon some of the concepts presented in the first edition of the present study (Schneiderman 1985). These reports present a very detailed analysis of maxillomandibular growth rotation in the rhesus monkey. They are based on a sample of ten males ranging from 1.5 to 5 years of age.

King (1990) has presented a detailed craniometric and cephalometric analysis of 308 rhesus monkeys from 0.5 to 24 years of age. The cross-sectional sample of monkeys used in this study are derived from the free-ranging population of Cayo Santiago. This study provides excellent data on sexual dimorphism and allometric changes as functions of age. Also, new information is presented on differential growth among the components of the palate as they contribute to sex differences in prognathism. Finally, this study presents the first craniofacial growth data from semiwild rhesus monkeys, making comparisons with laboratory-based data possible.

Rationale for Use of Bone Implants

A number of in vivo bone-marking procedures have been developed for studying changes in bone morphology. For longitudinal radiographic studies of bone growth and development, radiopaque metallic (tantalum) implants are much more suitable than vital stains. The small metallic implants used in this method provide a means for superimposing serial cephalograms of standard orientation accurately and objectively. Remodeling and displacement of the various components of the skull can be represented graphically and quantified with this approach.

Reliabilty of metallic bone implants has been a subject of much concern. Sarnat (1968) summarized the results of a series of experiments investigating the behavior of such implants in a variety of mammals, using cephalometric roentgenography in conjunction with standard dissection techniques. Local bone response to the implants and evidence of sutural apposition were evaluated macroscopically upon sacrifice of the animals. The radiographic and macroscopic evidence was found to be in agreement (cf. Selman and Sarnat 1953). Also, fewer than 4% of implants were lost from animals followed for periods ranging from 2 to 50 weeks in these experiments (Gans and Sarnat 1951; Robinson and Sarnat 1955; Roy and Sarnat 1956; Selman and Sarnat 1957). Furthermore, it was found that metallic implants produced only minimal local reaction. From these studies, Sarnat (1968) concluded that metallic implants are an informative means for investigating endochondral, appositional/resorptive, and sutural growth. These experimental studies also indicate that properly placed metallic implants provide an optimal means for superimposing head films taken in a standardized fashion.

Björk (1955a, 1955b, 1963) and Björk and Skieller (1972) developed and successfully applied the use of cephalographic radiography using radiopaque implants to the study of craniofacial growth in human children and adolescents. Björk (1963) demonstrated that when properly implanted in the mandible (i.e., placed in nonresorptive regions and at a distance from developing tooth buds), tantalum pins could be reliably used as fixed reference points for superimposition of serial radiographs, from which it was possible to describe metrically the magnitude and direction of condylar growth as well as appositional changes along the margins of the mandible. Comparable methods were developed for describing longitudinal changes in the human maxilla (Björk 1955a, 1966; Björk and Skieller 1977) and cranial base (Björk 1955b).

As an example of the profound utility of bone markers in studying craniofacial growth, Björk and Skieller (1972), expanding on earlier studies, found evidence of consistent patterns of jaw rotations in human children during growth. In this quantitative study, these investigators demonstrated that out of twenty-one circumpubertal children who were followed cephalometrically for six years, nineteen exhibited counterclockwise rotations (relative to a lateral cephalogram in which the face is oriented toward the right) of both jaws. From these findings they concluded that jaw rotations in this direction are a general feature of human facial development. They also reported that mandibular rotations were greater than maxillary rotations. Furthermore, they noted that peripheral remodeling of both jaws, as well as changes in the occlusal plane, tend to mask the rotational changes in the jaws. Additionally, they found significant correlations between rotational changes in the jaws and condylar growth. A major conclusion of this study is that compensatory remodeling and occlusal changes must occur to offset the effects of the rotations of the jaws if the face and occlusion are to develop normally.

Though very sophisticated in terms of the cepholometric analysis using implants, the studies of Björk (1955a, 1955b, 1963, 1969) and Björk and Skieller (1972, 1977) have been oriented toward case studies, that is, following individuals over time. These clinically oriented studies do not make a clear distinction between analytic methods useful for gaining general knowledge about growth processes from those useful for diagnosing and predicting the growth behavior of individuals. Although these studies have been oriented toward describing and predicting the growth behavior of children manifesting extreme facial morphologies (Solow, pers. comm.), this approach may be premature in light of the absence of good estimates of the parameters of *normal* craniofacial growth. To have biological credibility, descriptions of growth processes must be statistical, that is, population oriented, addressing the question of *variability* as well as central tendency. It is an appropriate clinical goal to describe and predict the growth behavior of the *individual*, but the *population* (or *subpopulation*) is the appropriate unit of description for biological processes.

Superimposition

Numerous cephalometric studies implicitly purport to describe how the components of the facial skeleton reposition relative to more central regions of the skull. It has been commonplace in the craniofacial biology literature, as well as in the clinical practice of orthodontics, to superimpose serial films on (1) an imaginary line connecting sella and nasion and registered on sella, or (2) the radiographic outlines of the anterior or middle cranial base. Each of these procedures is problematic for several reasons. The use of the sella-nasion line for superimposition involves a landmark on the nasomaxillary complex (nasion) and a landmark associated with the cranial base and brain (sella). Descriptions of how the components of the face (particularly the midface) reposition relative to the cranial base are confounded because the region being studied and the reference region are not completely independent. Furthermore, there is growth between these two landmarks, making their relationship to each other unstable.

The alternative procedures of superimposing on the roof of the orbits or sella turcica and/or other superior aspects of the sphenoid bone have other problems. These anatomical features partially remodel during growth in the human (Enlow 1982), in the rhesus (Duterloo and Enlow 1970), and in the pigtailed macaque (Sirianni and Van Ness 1978; Lestrel and Sirianni 1982), detracting from their suitability as stable elements upon which to superimpose. More importantly, superimpositions based *solely* on these remodeling anatomical features are largely subjective. Very subtle changes in the orientation of such superimpositions can significantly alter perceptions concerning the growth of the face. Simply put, without the aid of implants, an investigator can demonstrate virtually any type of growth, depending on subtle choices made during the superimposition process.

No set of landmarks, even of radiopaque implants, can produce a perfect superimposition to visualize the serial growth of the whole craniofacial complex. Because the growth of the skull is nonisomorphic and highly complex, the goal of finding such a superimposition technique is unrealistic, if not meaningless. However, there are at least three types of superimposition that are optimally suited for the description of both local and regional growth phenomena that provide biologically meaningful information.

To describe the repositioning of the components of the facial skeleton relative to the cranial base, it would be ideal to identify a small region within the cranial base that completes its development relatively early in life, is centrally located, and is related structurally to the neural components of the head. Such a region could be used as a reference relative to which changes in the more peripheral regions can be described.

Sphenoid Implants

The implants placed in the body of the sphenoid bone allow the replicatable superimposition on the middle cranial base, which meets the criteria described above (cf. Sirianni and Van Ness 1978). By superimposing on these implants one can readily describe the displacements of the components of the facial skeleton relative to the developmentally more mature cranial base, which, due to structural constraints imposed upon it by neural structures, maintains a more constant relationship to the brainstem and vertebral column (Enlow 1982). These implants also permit the description of changes in other parts of the neurocranium relative to the centrally located sphenoid bone.

Maxillary Implants

Implants placed in the body of the maxilla permit description of remodeling at the peripheral margins of the maxillary complex, that is, at the maxillary tuberosity, alveolar ridge, and facial aspect of the maxilla. Thus, by superimposing serial films or tracings on these implants, it is possible to identify those regions of the maxillary complex that have changed and those that remained constant in form. These implants are also useful in quantifying the displacement and rotation of the maxillary complex as a whole relative to the cranial base or mandible.

Mandibular Implants

Implants placed in the symphysis and body (corpus) of the mandible permit description of changes along the periphery of the mandible, for example, at the condyle, alveolar ridge, or margins of the ramus. They also permit the description of displacements and rotations of the mandible relative to the other elements of the skull.

2

MATERIALS AND METHODS

PRIMATE SAMPLES

Rhesus monkeys (*Macaca mulatta*) from two laboratory colonies were used in this study. The primary longitudinal sample consists of thirty-five animals, twenty males and fifteen females, studied at The University of Michigan between 1968 and 1984. Over this time period, numerous serial radiographs were taken on this sample, bringing the number of serial the films used in the present study to 374. All of the animals were unaltered in terms of experimental intervention and served as controls in various experimental studies conducted over the years by the Craniofacial Biology Group at The University of Michigan's Center for Human Growth and Development (CHGD).

The CHGD sample consists of several cohorts, and is technically a mixed-longitudinal sample. The distribution of animals is as follows: four of the males were studied for 8 or 9 years, and were first observed at 1 or 2 years of age. Nine males were followed for approximately 6 or 7 years, with their first observations made at ages ranging from 1 to 18 months after birth. The remaining seven males were followed for shorter periods of 2 years or less, most being observed for the first time at 2 to 3 years of age. The number of females in the sample was more limited, and therefore deemphasized in terms of the normative aspect of the study. Six of the females had 4 to 6 years of data, with their first observations at anywhere from 2 to 18 months of age. The remaining nine females were followed for shorter periods of 2 years or less, most being observed for the first time at 2 to 3 years of age. Although radiographs were taken on all animals at intervals of approximately 12 to 13 weeks, only films from every other observation were used for analysis. In other words, the standard interval for observation was approximately 26 weeks (one-half year) in length. The rational for this choice is discussed in the section subtitled "Variable Transformations."

The majority of the CHGD animals were wild-caught and were imported from India and Bangladesh as juveniles. The birthdates and, therefore, the chronological ages were known for the eleven monkeys in this sample that were born in captivity. They were all raised at the CHGD under standard laboratory conditions. They were fed a balanced diet consisting of Purina monkey chow, fruit, fruit beverages, and water *ad libitum*, and also received regular veterinary care from the Unit for Laboratory Animal Medicine. In terms of nutrition and health care, the conditions for normal growth were probably superior to those in the wild. However, the stress of living in captivity might have counterbalanced the positive effects of a controlled laboratory environment. The differences between captive and wild rhesus monkeys with regard to growth have been largely unknown, although findings on craniofacial growth in the free-ranging rhesus monkeys are now available (King 1990). We have recently conducted such a comparison with this semiwild population of Cayo Santiago, Puerto Rico (King and Schneiderman 1991). This preliminary report is summarized in the Discussion chapter.

The second sample is from the rhesus monkey colony at the Wisconsin Regional Primate Center (WRPC). This sample is comprised of a cross-sectional set of lateral cephalograms of forty-eight males and forty-three females, for whom birth dates are known. This structured sample was selected to represent adequately each half-year interval of the first 10 years of life, and was selected from a total sample of 320 films. These monkeys were raised under similar controlled laboratory conditions as those of the CHGD sample.

The primary focus of the study is maxillomandibular growth and development in the series of animals studied longitudinally. Therefore, most emphasis has been placed on measures of *maxillary and mandibular dimensions, positions, and remodeling* in the CHGD sample. On the other hand, the cross-sectional WRPC sample was used primarily to tackle the problem of age estimation (chapter 3). Thirty-seven measurements from the "Measurement Definitions" lists found later in this chapter were taken from the digitized tracings of the WRPC films. Represented in this set of variables are dimensions of the cranial vault and upper face, thicknesses of the bones of the vault, angulation of the cranial base, dimensions of the jaws, and positional measures of all the permanent molars and the deciduous second molars, as well as lengths of the roots of these teeth. These variables were chosen as indicators of (1) neurocranial, (2) splanchnocranial, and (3) dentitional maturation. It should be noted that some of the dental variables are limited to either juvenile or adult animals. Equations using these measures are based on a smaller subsample of monkeys representing only part of the total age span and therefore are more limited in their applicability (see Appendix B).

RADIOGRAPHIC CEPHALOMETRY

Radiopaque Bone Implants

Soon after the acquisition and quarantining of the CHGD animals, tantalum bone implants measuring 1.5 mm in length and 0.5 mm in diameter were surgically implanted percutaneously in the glabellar region, middle and posterior cranial base, and throughout the mandible and maxillary complex. Some animals also had markers implanted in various parts of the cranial vault. See Björk (1955a, 1955b, 1968), McNamara (1972), and Sirianni and Van Ness (1978) for details of the implanting technique.

Radiographic Technique

Serial lateral cephalograms of the CHGD animals were taken at regular intervals under general anesthesia. After administration of general anesthesia,[1] the head of the monkey was placed adjacent to the X-ray film and was centered using a specially designed cephalostat (McNamara 1972). This maintained the head in an identical, reproducible position at each radiographic session, ensuring the comparability of serial head films.

Lateral cephalograms were were taken with 60 to 70 kV with a tube to film distance of 134.62 cm (53 in). The distance between the midsagittal plane to the film plate varied with the width of the animal's head but ranged between 35 and 65 mm. All cephalograms were made with a consistent enlargement factor using Kodak XG-1 XOMAT or type M Industrial film. At each session, a film was taken with the teeth maximally interdigitated (a rubber band around the muzzle of the animal was used to hold this jaw position constant) and another film was taken with the jaw opened about 3 cm interincisally by a Plexiglas rod. This latter film provided a better view of the condylar and coronoid regions of the mandible. Tracings of the films were made on 0.003 mm matte acetate. A narrow pencil lead (0.3 mm HB) was used for making the tracing.

The animals in the cross-sectional WRPC data set did not have bone and muscle implants. Therefore, the WRPC films were traced and digitized using the same scheme of landmarks and reference points used for the CHGD animals, except for

[1] Phencyclidine HCl [Sernylan], a dissociative agent, and pentobarbitol Na were used from 1968 to 1978. Ketamine HCl, a dissociative agent, and xylazine HCl, a muscle relaxant and analgesic, were used from 1978 to the present.

the last fourteen points which represent the implants and constructed points (see the section, "Tracing Protocol and Point Definitions").

Computer Techniques

Data were collected in the Craniofacial Biology Laboratory at the Department of Orthodontics of The University of Michigan using an integrated system consisting of a microcomputer, digitizer, digital plotter, and printer, as well as custom software (see Appendix A for details regarding specific software and hardware used). The tracings of the cephalograms were digitized sequentially and hard copy plots of each of the skulls were generated to verify the accuracy of the stored data with regard to the original tracings and radiographs. Composite plots of serial films superimposed on appropriate bone implants were produced to verify the accuracy of the superimpositions as well as the *change* data that issues from such superimpositions. The digitizing and plotting were accomplished using the *Ceph-Master*[2] software system designed especially for handling cephalometric data. Chronological ages were estimated using custom programs on the microcomputer and the statistical package *MIDAS* (1976) on the mainframe computer at The University of Michigan. The digitized coordinate data were uploaded from the microcomputer to the mainframe for further data manipulation, including the merging of age estimations into the coordinate data set. The coordinate data were analyzed using two statistical packages (*MIDAS* [1976] and *SAS* [1982a, 1982b]) on the mainframe. *SAS* was used for developing and executing special statistical procedures for analyzing longitudinal data. *MIDAS* permitted direct computation of changes between successive films. Cephalometric distances and angles were calcu-

[2] Ceph-Master © 1984–1987 by Trilobyte Software, Inc., 15894 Northville Rd., Plymouth, MI 48170.

lated using the *EXTRACT* program written by Richard Miller that is available through the CHGD on the mainframe computer.

Tracing Technique

A tracing scheme was developed to represent the shapes and spatial relationships of the components of the craniofacial complex using as few points as possible. These points, when connected by line segments, constitute a geometric representation or model of the skull and dentition. Eighty-three points were required to portray the curvatures and inflections in the radiographic image of the head that give the osseous elements their characteristic shapes. An additional seventy-nine points were required to represent the position of the deciduous and permanent teeth, and fourteen more points were used to represent the positions of the bone and muscle implants.

The choice of landmarks/points and the overall tracing strategy were designed to maximize biological information. Many conventional landmarks as well as measurements were avoided because they tend to confound the accurate description of interrelationships of growth between the various components of the craniofacial complex (cf. Moyers and Bookstein 1979). Discrete osseous components were traced independently of others; for example, when tracing the mandible on a series of longitudinal films of an animal, the homology of landmarks on this structure is based on information in the mandible alone. The points are not defined with reference to cranial base or maxillary landmarks. An explicit example of where this strategy differs from a more conventional approach is the elimination of landmarks such as *articulare*, a point on the mandible defined by its position relative to the cranial base. By defining landmarks principally on the basis of local features, one minimizes correlations that have their basis in tracing technique rather than in biological reality.

The landmarks chosen to be digitized on the lateral cephalogram do not represent perfectly the midsagittal section of the skull. Rather, they are a compromise between representing the hard structures of the head in a consistent manner while maximizing efficiency and manageability in the tracing and digitizing process. The various rules and conventions used permit the tackling of the formidable task of adequately representing the major components of the skull and dentition with some degree of replicability (see section on Data Verification and Error Analysis). These conventions are being described in detail to inform those who choose to use this normative data of the inherent biases it contains. This information should allow these investigators to make necessary compensations or adjustments.

Since this project focuses on *changes* in dimensions and angles rather than on absolute values, intra-individual consistency was emphasized rather than interindividual consistency in the identification of homologous landmarks. For example, when double images were present on radiographs, the more posterior of the two outlines was generally chosen.

Radiographic evidence of morphological change on a film relative to the prior film of the series was recorded only when it was obvious and indisputable when evaluated relative to the entire series of films taken of that animal. This conservative guideline may have introduced a bias toward underestimating change. It may also tend to make growth curves for the individual animals slightly less smooth than they actually are.

Tracing Protocol and Point Definitions

The following is a detailed description of the points that were digitized in this project. Traditional anthropometric and cephalometric landmarks are indicated in capital letters. The corresponding abbreviations are from Riolo and coworkers (1974) when available. Other traditional craniometric points not defined in this source are derived from standard cephalometric and anthropometric references (Martin 1928; Krogman and Sassouni 1957; Comas 1960). The named landmarks in the rhesus monkey have been *adapted* from their human homologues; they are *not* identical. Also, craniometric points traditionally taken on three-dimensional skulls have been redefined for the two-dimensional cephalogram.

The choice of outlines and the definition of some points are based on tracing experience gained from previous cephalometric studies (Schneiderman and Carlson 1981, 1983, 1985; Carlson, Ellis, et al. 1982; Carlson and Schneiderman 1983) that lead to the choice of contours and points that can be located most consistently and accurately. Experience indicates that this approach permits the accurate representation of subtle longitudinal changes within animals, though it may not be ideal for cross-sectional data sets.

Craniofacial Skeleton
(Figs. 2.1–2.3)]

MIDDLE AND POSTERIOR CRANIAL BASE (1–11)

1–3. Curving contour of anterior aspect of greater wing of sphenoid bone. If image was double on the initial film(s), the more posterior image was traced on this film and throughout the series.

1. Distinctive beginning of the curvature located a few mm inferior to the floor of the anterior cranial fossa.
2. Contour begins to curve posteriorly at this point.
3. Posterior end of curvature. Near the junction of the ethmoid, vomer, and palatine bones.
4. BASION, BA. Most anterior point on foramen magnum. This point, though difficult to visualize, was located consistently by tracing the occipital condyles and noting the relationship between these structures and basion. Also, by tracing the spinal canal superiorly, it was possible to identify the most posterior possible position that basion could occupy. A similar relation pertains to opisthion (see below) and the posterior aspect the spinal

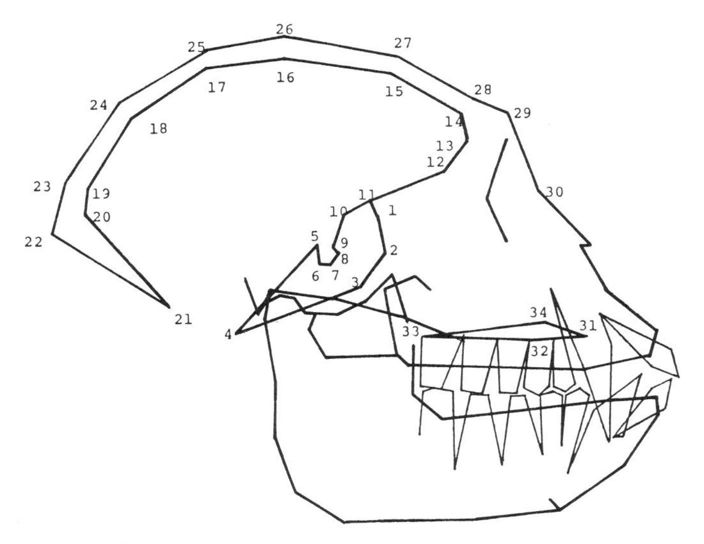

FIGURE 2.1 Outlines used to represent the skull and dentition in tracings and plots. Cranial vault and hard palate landmarks are indicated by their numbers in the digitizing scheme.

canal. The linear distance between basion and opisthion is approximately 15 mm.

5. Anterosuperior tip of posterior clinoid process.

6–8. Equally spaced points between posterior and anterior clinoid processes which represent the curvature of sella turcica.

9. Posterosuperior tip of anterior clinoid process.

10. Most superior point on rim of optic foramen.

11. Point at which the radiographic outlines of the floor of the anterior cranial fossa (in the midsagittal plane) and the orbital roof (in a parasagittal plane, approximately 5 mm lateral to midline) intersect.

ENDOCRANIAL OUTLINE (12–21)

12. Approximate midpoint along the floor of the anterior cranial fossa where it is crossed by the most posterior contour of the postorbital bar(s).

13. Most anterior point of brain case indicated where the

endocranial outline is crossed by the most anterior extent of the orbital roof.

14–20. These endocranial points were placed directly across (i.e., at the shortest possible distance) from their ectocranial counterparts (see below), some of which are formally defined.

21. OPISTHION, OP. Most posterior point on rim of foramen magnum (see basion above, for determination of its location).

ECTOCRANIAL OUTLINE AND UPPER FACE (22–30)

22. INION, I. Most posterior tip of external occipital protuberance.

23. LAMBDA, L. Lambdoidal suture between occipital and parietal bones. In the event that this or any of the other vault sutures was not visible on any of the films of a series, these points were placed at subequal intervals to best represent the curvature of the vault.

24, 25, 27. Equally spaced constructed points between the landmarks lambda, bregma, and ophryon.

26. BREGMA, BR. Bregmatic suture between frontal and parietal bones (a.k.a. frontal or coronal suture).

28. OPHYRON, ON. Greatest depth of concavity (sulcus) posterior to the supraorbital torus. In immature animals, in which a well-developed sulcus had not emerged, its presumptive site was placed where the frontal bone was thinnest.

29. GLABELLA, GB. Most anterosuperior point on supraorbital torus.

30. NASION, N. Point of greatest concavity along anterior contour of nasal bone. It is also the approximate midpoint between glabella and rhinion (see below).

HARD PALATE (PALATINE PORTIONS OF THE MAXILLA AND PALATINE BONES) (31–34)

31. Anterior end of hard palate, just posterior to incisive foramen.

32. Greatest concavity along inferior (oral) aspect of palate.
33. POSTERIOR NASAL SPINE, PNS. Posterior end of palate.
34. Greatest convexity along superior (nasal) aspect of palate.

ORBIT (35–37) (FIG. 2.2)

35. Most superior point on orbital rim, near supraorbital notch or foramen. The locating of this point was aided by noting the junction between the orbital roof and facial aspect of the frontal bone.
36. Most posterior point along orbital rim, approximately at the midpoint between points 35 and 37.
37. ORBITALE, OR. Most inferior point on orbital rim. Locating this landmark was aided by noting the junction of the floor of the orbit and the facial aspect of the maxilla.

MAXILLARY COMPLEX AND MIDFACE (38–51)

38. RHINION, RH. Anterior tip of nasal bones.
39. Most superior point on nasal aperture, located a few mm lateral of midsagittal plane.
40. Midpoint along anterior maxillary contour between points 39 and 41.
41. Tip of the thin wedge of alveolar bone which appears to overhang the central incisor in the lateral radiographic view. In the frontal view of the skull, this point is actually the highest point along the arc of bone where the tooth emerges from the alveolus.
42. SUPRADENTALE, SD. Most anterior point along the maxillary alveolar ridge. Difficult to visualize in young animals, the location of this point was interpolated by extending the curvature of the maxillary alveolar ridge anteriorly and the curvature of the anterior maxillary contour inferiorly and noting their intersection. It is usually overlaid by the image of the pulp cavity of the central incisor.
43. Point along maxillary alveolar ridge where it is crossed by the distal edge of the canine tooth.

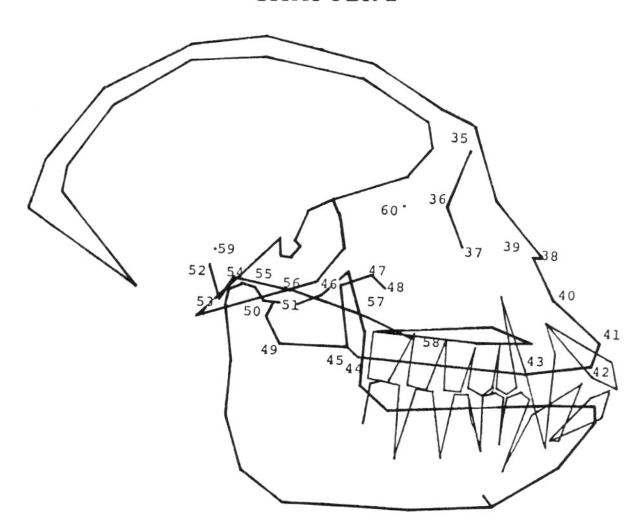

FIGURE 2.2 Facial mask and temporomandibular joint landmarks.

44. Midpoint along the non-tooth-bearing portion of the maxillary alveolar ridge distal to the last molar.
45. MAXILLOPALATINE JUNCTION, MXPJ (original landmark). The inferior end of the junction between the maxillary tuberosity and the pyramidal process of the palatine bone. It is located at the top of a well-defined notch between these two osseous processes.
46. PTERYGOMAXILLARY FISSURE, INFERIOR, PTMXFI. Point at which the pterygomaxillary fissure opens superiorly into the pterygopalatine fossa, an obliquely oriented teardrop-shaped aperture. This point is usually overlaid by the image of the zygomatic portion of the zygomatic arch.
47. Most superior point on pterygopalatine fossa.
48. Most anterior point on pterygopalatine fossa.

LATERAL PTERYGOID PLATE OF SPHENOID BONE (49–51)

49. Posterinferior corner of plate. This point was constructed by bisecting the angle between lines representing the inferior and posterior borders of the plate.

50. Most inferior junction between the plate and temporal bone. At the apex of a notch between these structures.
51. Most inferior point on foramen ovale.

TEMPORAL COMPONENT OF TMJ AND ZYGOMATIC ARCH (52–58)

52. PORION, PO. Anatomic porion. Most superior point on external auditory meatus.
53. Most inferior point on postglenoid spine.
54. Most superior point on the mandibular (glenoid) fossa. This point was consistently identified on cephalograms as the apex of the least radiopaque region of the joint. On the actual skull it is located at the lateral most aspect of the temporal component of the joint. This radiographic point is slightly inferior (< 1 mm) to the more medially located roof of the joint.
55. Most inferior point on the articular eminence (tubercle).
56. Point at greatest depth of concavity along inferior aspect of the zygomatic arch. It is near the suture between the temporal and zygomatic portions of the zygomatic arch.
57. Midpoint between points 56 and 58 along the arch.
58. Root of zygomatic arch. Near inferior end of zygomaticomaxillary suture. It is in approximately the same transverse plane as the hard palate.

ATTACHMENTS OF TEMPORALIS MUSCLE (59–60)

59. Point along suprameatal crest directly superior to porion.
60. Location on posterior aspect of postorbital bar, as seen on the dry skull, where various ridges (attachments for temporalis m.) converge. Just posterior to the greatest concavity in the image of the orbital rim.

MANDIBLE (61–83) (FIG. 2.3)

61. Most anterior point along anterior border of the mandible. Analogous in location to point 41 on the maxilla.

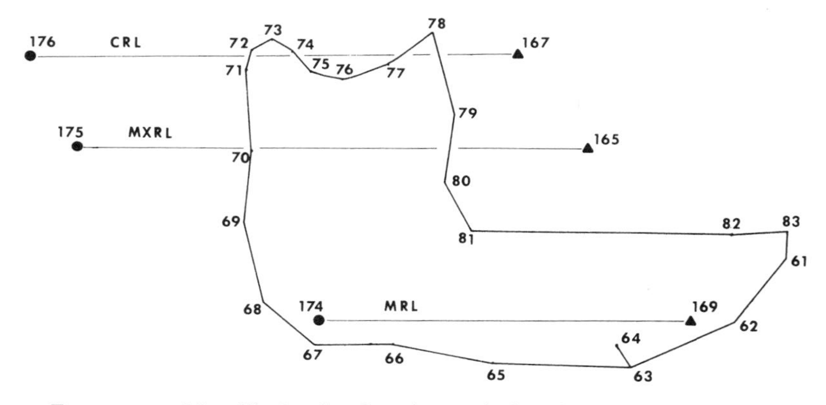

FIGURE 2.3 Mandibular landmarks and the three reference lines used for superimposition. CRL is the Cranial Reference Line. MXRL is the Maxillary Reference Line, and MRL is the Mandibular Reference Line. The triangles are bone implants and the circles are constructed points.

62. POGONION, PG. Point along anterior contour of the mandible that lies midway between points 61 and 63 (menton). It is also approximately at the greatest convexity of this contour.
63. MENTON, ME. The point of junction between the most inferior end of the symphyseal outline and the antero-inferior border of the mandible.
64. Most posterior point along the lingual contour of the symphysis. Posterior tip of mental spine (genial tubercle).
65. Point of greatest convexity along the segment of the inferior border of the corpus anterior to premasseteric notch. This point was located by running a line along the full length of the inferior border of the mandible that is tangent at this point and another point at the gonial region (point 67).
66. Premasseteric (antegonial) notch. Point of greatest concavity of inferior border of mandible located approximately at the most anterior part of the masseteric insertion.

67. Point of greatest convexity along segment of inferior border of mandible that is posterior to premasseteric notch. Also see point 65.
68. GONION, GO. Midpoint of mandibular (gonial) angle. Located on initial film of a series by bisecting the angle between the bitangent line along the inferior border of the mandible (described for point 65 above) and a similarly formed bitangent line along the posterior border of the mandibular ramus.
69. Point of greatest convexity along the lower part of the posterior border of the ramus.
70. Point of greatest concavity along the posterior border of the ramus, which is approximately the midpoint between points 69 and 71.
71. POSTERIOR CONDYLION, PCO. Most posterior point on condyle, where it begins to curve forward. It also approximates the beginning of the articular surface of the condyle (cf. Elgoyen et al. 1972).
72. CONDYLION, CO. Most posterosuperior point along the condyle. It is located by bisecting the angle between the bitangent line along the posterior border of the ramus and its perpendicular line, which is tangent to the condyle at its superiormost point, superior condylion. Condylion is located approximately at the midpoint of the articular surface of the condyle.
73. SUPERIOR CONDYLION, SCO. Most superior point on the condyle. (See point 72 for its determination.)
74–75. Equally spaced points between points 73 (superior condylion) and 76.
76. Point of greatest concavity along mandibular (sigmoid) notch.
77. Midpoint between points 76 and 78.
78. Most superior tip of coronoid process. Difficult to visualize on many films, this landmark is located near the inferior border of the ethmoid bone (point 3). With this in mind, its position was interpolated by tracing the an-

terior margin of the ramus upwards to this region; this was best done on a film with the mandible in the open position.

79. Point of greatest convexity along anterior margin of ramus. Approximately at the midpoint between points 78 and 80.
80. Point of intersection between functional occlusal plane and anterior margin of ramus. Functional occlusal plane must be defined with the teeth of both jaws fully interdigitated.
81. Most posterior point along mandibular alveolar ridge. It was defined as the radiographic intersection of the images of the most inferior extent of the anterior ramal margin (oblique line) and the alveolar ridge.
82. Point of intersection between mandibular alveolar ridge and distal edge of mandibular canine tooth.
83. INFRADENTALE, ID. Most anterior point on alveolar ridge. It was located analogously to its maxillary counterpart, supradentale, point 42 (see above).

Dentition and Implants (Figs. 2.4–2.6)

Terminology for dental landmarks is consistent with that in *Dental Anatomy and Occlusion* (Kraus et al. 1969), wherever possible.

Any given cephalogram does not possess the full complement of both the deciduous and permanent dentition; one set of teeth, or a significant portion thereof, was therefore entered as missing data. When the teeth appeared as a double image, as was the case for most films, the more posterior image of all of the teeth was chosen, except the incisors. The more anterior outline of the central incisors was used. The cusps and roots were defined radiographically, not strictly anatomically. In the rhesus monkey, the mesiobuccal and distobuccal cusps tend to be slightly more posteriorly located than their lingual counterparts on both the mandibular and maxillary molars. Therefore, the radiographic designation of "distal cusp of the second maxillary molar" would most likely correspond anatomically to the distobuccal cusp of the more posteriorly located tooth of the two maxillary second molars.

On teeth in which the root tip is not fully formed and therefore radiographically indistinct, the following procedure was used. A line was drawn between the anterior and most posterior images of the most apical aspect of the tooth. The root tip point was designated as the midpoint of this line. In immature teeth in which only the crown was formed, the root tip point was entered as missing data.

On each tooth the landmarks were ordered in a clockwise fashion, starting with the most anterior landmark. All incisors, canines, premolars, and mandibular third molars were each represented by four landmarks. All of the other molars were represented by only three landmarks each.

The definitions of landmarks on tooth crowns (e.g., #88, "Point of greatest convexity along anterior [mesial] aspect of the maxillary canine") concern intrinsic features of the teeth, and are therefore unaffected by their position or extent of eruption.

MAXILLARY DECIDUOUS DENTITION (84–97) (FIG. 2.4)

84–87. Maxillary central incisor. The more anterior of the two was traced if both were visible.
84. Point of greatest convexity of labial aspect.
85. Incisal tip.
86. Point of inflexion along lingual aspect.
87. Root apex.
88–91. Maxillary canine.
88. Point of greatest convexity along anterior (mesial) aspect.
89. Cusp apex.
90. Point of greatest concavity along posterior (distal) aspect.

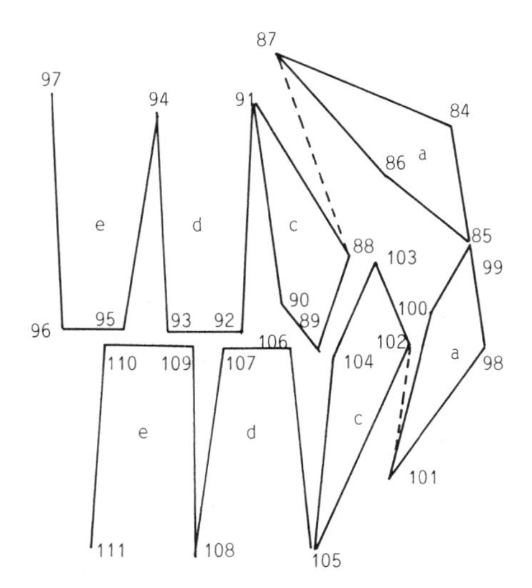

FIGURE 2.4 Deciduous dentition landmarks.

91. Root apex.
92–94. First maxillary molar.
92. Mesial cusp tip.
93. Distal cusp tip.
94. Distal root tip.
95–97. Second maxillary molar.
95. Mesial cusp tip.
96. Distal cusp tip.
97. Distal root tip.

MANDIBULAR DECIDUOUS DENTITION (98–111)

All points on these teeth were defined analogously to those on their maxillary antagonists.

MAXILLARY PERMANENT DENTITION (112–136) (FIG. 2.5)

112–115. Maxillary central incisor. The more anterior-most of the two was traced. Point definitions are the same as for the deciduous central incisor (see points 84–87).
116–119. Maxillary canine. Same as deciduous canine (see

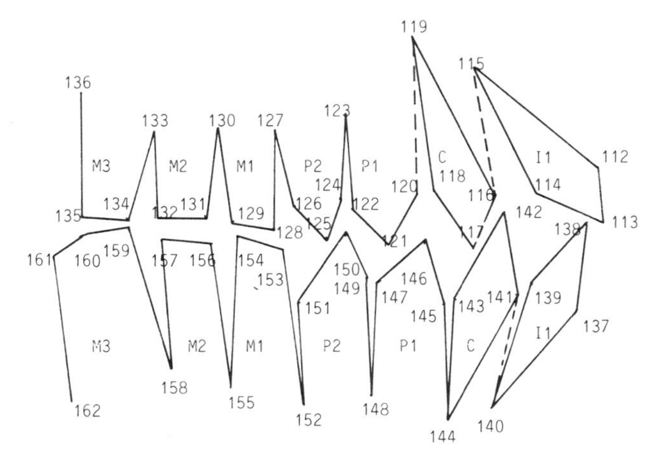

FIGURE 2.5 Permanent dentition landmarks.

points 88–91). Note that these teeth are quite large and take a long time to develop fully, particularly in the male. Because of this, it was necessary to refer to a fully adult film in a series to locate correctly points 116 and 118, which were typically constant in position relative to the tooth cusp throughout the animal's life.

120–123. Maxillary first premolar (quite different in shape from the human counterpart).

120. Point of inflexion where anterior interproximal aspect and occlusal aspect meet.

121. Cusp apex.

122. Point of inflexion where posterior interproximal aspect and occlusal aspect meet.

123. Distal root tip.

124–127. Maxillary second premolar. Same as maxillary first premolar (points 120–123).

128–130. Maxillary first molar. Landmarks are the same as for the deciduous maxillary first molar (points 92–94).

131–133. Maxillary second molar. Same as maxillary first molars, deciduous and permanent (points 95–97 and 128–130, respectively).

134–136. Maxillary third molar. Same as maxillary first and second permanent molars (points 128–130 and 131–133).

MANDIBULAR PERMANENT DENTITION (137–162)

All landmarks for these teeth were defined analogously to their maxillary antagonists except the third molar.

159–162. Mandibular third molar.
159. Mesial cusp tip.
160. Distal cusp tip.
161. Occlusal tip of talonid.
162. Distal root tip.

BONE MARKERS (163–171) (FIG. 2.6)

These are radiopaque tantalum implants (measuring 1.5 × 0.5 mm), surgically implanted subperiosteally in various regions of the skull.

163. In frontal bone somewhat superior to glabella.
164. In premaxilla. As the premaxillary suture closes with maturity, the position of this marker, with respect to the others in the maxillary complex, stabilizes.
165. In body of maxilla, superior to first or second molar.
166. In the region of the zygomaxillary suture. Usually in the zygomatic bone.
167. In body of the sphenoid bone, inferior to sella turcica. One of the more anterior implants in this region was chosen.
168. In the basioccipital portion of the occipital bone posterior to the spheno-occipital synchondrosis.
169. In symphyseal region of mandible.
170. In corpus of mandible.
171. In ramus of mandible.

MUSCLE MARKERS (172–173)

172. In distal region of right masseter. Markers that were placed in this side are short and thick. A marker as close to gonion as possible was chosen.

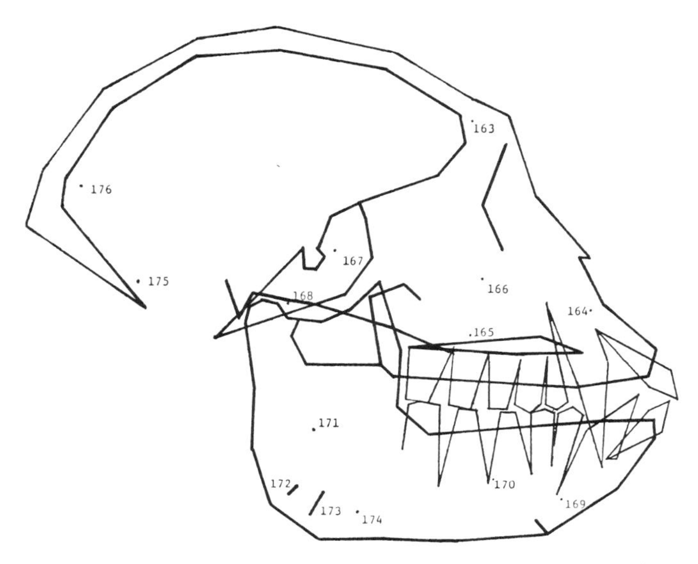

FIGURE 2.6 Locations of radiopaque bone implants (163–171), muscle implants (172–173), and constructed points (174–176). The constructed points were used to define the reference lines involved in superimposition (see section entitled "Reference Line Construction").

173. In distal region of left masseter. Markers are long and thin. A marker close to gonion was chosen.

CONSTRUCTED POINTS (174–176)

These points were constructed from the positions of the various bone implants and were required for the three superimposition schemes. They were marked on the films themselves at radiopaque locations. They are described in greater detail in the section below on Reference Line Construction (see also figs. 2.3 and 2.6).

174. Point at the posterior end of the mandibular reference line (MRL), which is parallel to the original occlusal plane and passes through the symphyseal bone implant (point 169).

175. Point at the posterior end of the maxillary reference line (MXRL). Constructed analogously to point 174, it is

on the line which is parallel to the original occlusal plane and passes through the maxillary body implant (point 165).

176. Point at posterior end of cranial reference line (CRL). This line was defined on the first (or early) film of a series. It is a line that passes through the sphenoid implant (point 167) and is parallel to the original occlusal plane. This constructed point was placed at the posterior end of the line near the endocranial outline.

CEPHALOMETRIC MEASUREMENTS

Types of Variables

Dimensional variables are simple linear distances between two defined landmarks. These scalars describe the major dimensions of the various osseous units of the skull. Linear distances were also used to describe tooth length and position. An example of a dimensional variable is *maxillary length* (MXLN). *Serial changes* in these dimensions for half-year intervals were also calculated, and are referred to, for example, as *change in maxillary length* (MXLN1). Note that the code name is the same as that of the dimensional variable itself, except it is followed with a "1," to indicate a lag of one. These and all other nonangular measures are expressed in mm.

Relational variables were used to describe the positions of the various osseous units relative to each other. These were also used to describe dental position. Relational measures fall into two categories:

1. *Translational* variables describe the translation of a portion of an osseous unit, as identified by a bone implant, relative to another (reference) region of the skull, as represented by its implants. These are defined as horizontal (X) and vertical (Y) displacements (scalars), where the horizontal plane is arbitrarily defined as parallel to the *original occlusal plane*, and the vertical as its perpendicular through an implant in the reference region (see section "Reference Line Construction" for details of this

method). These are all *change* variables, that is, differences or "deltas" between successive observations within individuals generated by the "relative time" processing capability of MIDAS (Fox and Guire 1976). An example of a translational variable is the *horizontal displacement of the mandible relative to the cranial base implants*, XMNDSPL1.

It should be noted that the partitioning of resultant vectors of displacement or growth into horizontal and vertical components is no more than a useful convention. These orthogonal scalars are easy to manipulate and are readily analyzed statistically, unlike their vectorial counterparts.

2. *Rotational* variables were used to describe the angular relationship between osseous units, or between osseous units or teeth and reference planes. The former type was determined by calculating the angle between the *reference lines* of the various units. Since the reference lines are defined by bone implants, only the *change* variables based upon these angles are meaningful. An example is the *change in angle between the mandibular and maxillary reference lines* (MRLMXRL1). An example of the latter type is the *change in angle between the maxillary reference line and the occlusal plane* (MXRLOCPL1). These are all expressed in degrees counted in a clockwise fashion in an arc beginning at the horizontal plane. Thus a change value with a negative sign signifies counterclockwise rotation. Anatomically, the terms "clockwise" and "counterclockwise" are used with reference to lateral cephalograms in which the face is oriented toward the right.

Remodeling or *local change* variables describe the translation of landmarks relative to the osseous units upon which they reside. Computationally, they are the displacements of the landmarks relative to the appropriate reference line (i.e., MXRL or MRL). More specifically, they are the serial differences in the orthogonal distances from the landmark in question to the origin of the coordinate system based on bone implants in the unremodeled portion of the body of the osseous unit. An example is the *vertical displacement of condylion* relative to the mandibular symphysis and body implants (YCO1).

Though some are cumbersome, the pneumonic variable

names have been included throughout this book for two reasons: (1) for the reader concerned with technical detail, they permit cross-referencing the discussion of these measurements in the text with the figures and tables; (2) they are also used to differentiate clearly between the *measurements* and the *phenomena* they are intended to represent. For instance, the variable YMXALVR1, *vertical change at the maxillary alveolar ridge*, is the author's attempt to represent the vertical growth in this region. Thus, it will be clear to the reader when the specific results of this study ("the data") are being presented, and when biological phenomena, as inferred from these measures, are being discussed.

Variable Transformations

Since the length of time elapsing between successive radiographs is variable throughout this data set (ranging from 3 to 70 weeks), it was necessary to standardize the change variables for a unit time interval. A half-year interval (26 weeks) was chosen as the unit rather than quarter- or full-year intervals. This choice was made because it maximized the number of subjects per observation (there were more subjects for any given half-year period than for each of the corresponding quarter-year intervals) while permitting the exhibition of subtle trends that were obscured by lumping observations into larger intervals.

In an earlier study (Schneiderman and Kowalski 1985) it was shown that linear equations were adequate for representing the growth curves of some jaw dimensions over periods of time less than 2.5 years. Therefore, it is reasonable to standardize variables for a unit time interval since their behavior over the relatively short periods of time elapsing between films is approximately linear. Some information may be lost by standardizing the data in this fashion, but it allows the grouping of the observations in a manner that makes statistical comparisons possible.

The vectoral equivalents of the orthogonal pairs of standardized change variables were calculated using trigonometric functions of a general triangle. The *directions* of these vectors are named in the same way as the corresponding orthogonal change variables except that they are suffixed with the letter "D" rather than the number "1." These are presented to make the directional information contained in the orthogonal change variables more salient.

Data Verification and Error Analyses

A number of procedures were used to verify the accuracy and quantify the amount of error in this data set. These procedures, designed to maximize the reliability of these data, were performed at various phases of the development of the data set.

Data Acquisition Phase

Single scale plots of each of the digitized tracings were produced using a high-resolution digital plotter (HP7475A). Each of the plots was examined visually and metrically where necessary. The scheme used to plot the coordinate data connects the points using color so that most tracing or digitizing errors were made obvious.

Composite plots of the digitized data were made using the same plotter. These were used to visualize serial change within the various parts of the skull (i.e., mandible, maxilla, and cranial base) by superimposing on the appropriate implants. These plots indicate graphically where change has occurred within individuals, given that the superimposition is correct. Thus they permit scrutinization of the criteria used for superimposition itself, so that depictions of biologically impossible or unlikely change can be identified and corrected. Examination of the original films and tracing were required in some series of films to distinguish between real and artifactual change.

These composite plots were also used to gain a comprehensive picture of morphological change within and among the various regions of the craniofacial complex that is unavailable when dealing with numbers alone. In addition, these composite plots were useful in verifying the accuracy of the quantitative data.

Tracing error was assessed by tracing a single radiograph six times and digitizing each of the tracings. The full complement of variables were generated for these replications and analyzed statistically using *MIDAS*. The overall average standard deviation for all the linear measurements of the maxilla that are based on landmarks alone is 0.241 mm, ranging from 0.169 to 0.312 mm. For those maxillary variables that use the radiopaque bone implants, the standard deviations average 0.217 mm and range from 0.056 to 0.342 mm. Standard deviations for all angular measures average 0.404 degrees and range from 0.160 to 0.952 degrees. For the mandibular measures using landmarks only, the standard deviations average 0.288 mm and range from 0.223 to 0.386 mm. For those mandibular measures using bone implants, the standard deviations average 0.259 mm and range from 0.165 to 0.421 mm.

The differences among these multiple tracings for each of the thirty-four variables were also analyzed. Among truly serial films these differences should indicate morphological change; in multiple tracings of the same film they should approximate zero. All of the linear change variables average 0.0194 mm. The averages for each of the linear variables range from 0.003 to 0.138 mm. The angular change variables range from 0.0029 to 0.528 degrees and average 0.237 degrees.

For any particular pair of films, the maximum value for a change variable is 1.039 mm and the minimum is 0.0931 mm. In addition to reflecting error or inconsistency in tracing the landmarks, these deviations also reflect error in the criteria for superimposition, that is, identification of the implants and constructed points. While some of these individual deviations are high, it is important to note that they are readily swamped out in this sample of six films by a majority of "correct" tracings. For instance, for the measurement of *ramus height*, RMHT, which exhibits the maximum value of 1.039 mm, the average for all the films for this variable is 0.0224.

Digitizing error was assessed by digitizing a single tracing six times and analyzing the results statistically with *MIDAS*. Means for each of the thirty-four basic variables were generated. For the linear measurements, the average standard deviation was 0.224 mm for the linear measurements and 0.239 degrees for the angular measurements. The standard deviations of the linear variables range from 0.0768 mm to 0.310 mm. The standard deviations for the angular measurements range from 0.152 degrees (*rate of maxillary rotation relative to* CRL, MXRLCRL1) to 0.340 degrees.

As an additional test of the accuracy involved in the digitizing process as well as in the superimposition algorithm used by the computer software (see Appendix A), the serial differences among the films for the thirty-four measurements were analyzed. The linear change variables average 0.00676 mm and range from 0.000922 mm to 0.104 mm. The angular change variables average 0.0492 degrees and range from 0.0137 to 0.0738 degrees. Thus, the error involved in the digitization and computerized superimposition processes is minuscule.

Verification of Descriptive Statistics

Each cephalometric measurement was examined for normality and outliers using *MIDAS*. Descriptive statistics (mean, standard deviation, minimum and maximum, and measures of skewness and kurtosis) and histograms were generated for the entire data set (i.e., all time periods treated together). The criteria of normality served to establish which variables were suitable for analysis with parametric statistics, as well as to identify those that may have been calculated incorrectly or have outliers.

Outliers were identified as follows. Those variables having high standard deviations (i.e., greater than 2 mm for change variables) were examined closely. The data set was stratified into progressively smaller groups by sex and age group to find the specific cases responsible for any excessive variability. Measurements were made manually on the individual and composite plots to clarify whether deviant values were real or artifact. In some ambiguous situations it was necessary to examine and compare the original radiographs, tracings, and plots to locate the source of the deviant values.

Measurement Definitions

Consult the listing of point definitions for complete landmark definitions. These measurements are illustrated in figures 2.7–2.9.

Maxillary Measurements
(Fig. 2.7)

XSD1 and YSD1. Horizontal and vertical changes at *supradentale* (42) relative to the maxillary reference line (MXRL).

XMXPJ1 and YMXPJ1. Horizontal and vertical changes at *maxillopalatine junction* point (45) relative to the maxillary reference line (MXRL).

XPTMXFI1 and YPTMXFI1. Horizontal and vertical changes at the *pterygomaxillary fissure* point (46) relative to the maxillary reference line (MXRL).

YMXALVR1. Vertical changes at the *maxillary alveolar ridge* in the region of the canine tooth (at point 42) relative to the maxillary reference line (MXRL).

MXLN. *Maxillary length* measured from the most anterior point of the maxilla (41) to the maxillopalatine junction point (45).

MXHT. *Maxillary height* in the posterior region as measured from a point at the distal end of the alveolar ridge (44) to the point at the superior aspect of the pterygopalatine fossa (47).

PALLN. Hard *palate length.* Measured from the incisive canal (31) to the posterior nasal spine (33). (Taken for age estimation only.)

XMXDSPL1 and YMXDSPL1. Horizontal and vertical *displacement* of the *maxillary complex* relative to the cranial base. Operationally these are the orthogonal displacements of

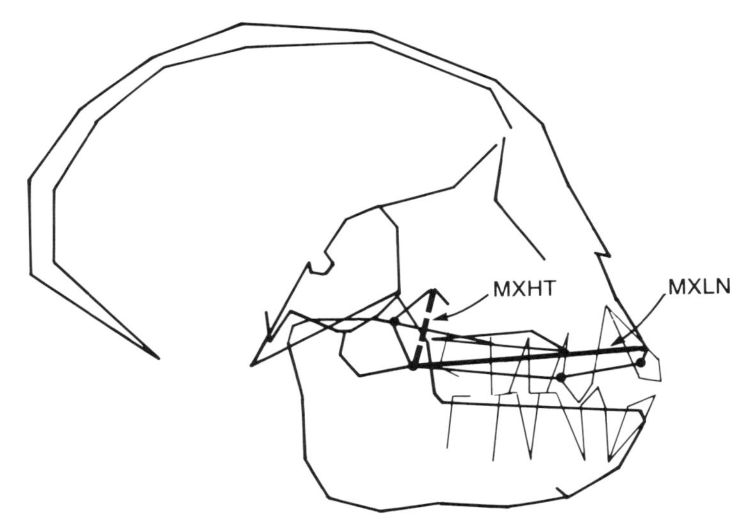

FIGURE 2.7 Maxillary dimensions and landmarks shown on a plot of a young male. Remodeling at each of the points indicated has been measured.

the maxillary bone implant (165) relative to the cranial reference line/coordinate system.

MXRLCRL1. Change in *angle* between maxillary body (as represented by maxillary reference line) and cranial base (as represented by cranial reference line).

MXRLOCP1. Change in *angle* between maxillary body and functional occlusal plane (Krogman and Sassouni 1957). The functional occlusal plane was determined by inspection. For actual computations, the maxillary occlusal plane was represented by the point at which it intersects the anterior margin of the ramus (80) and the mesial cusp of the first molar (point 92 in subadults, 128 in adults).

Mandibular Measurements
(Fig. 2.8)

XID1 and YID1. Horizontal and vertical changes at *infradentale* (61) relative to the mandibular reference line (MRL).

XMN1 and YMN1. Horizontal and vertical changes in the po-

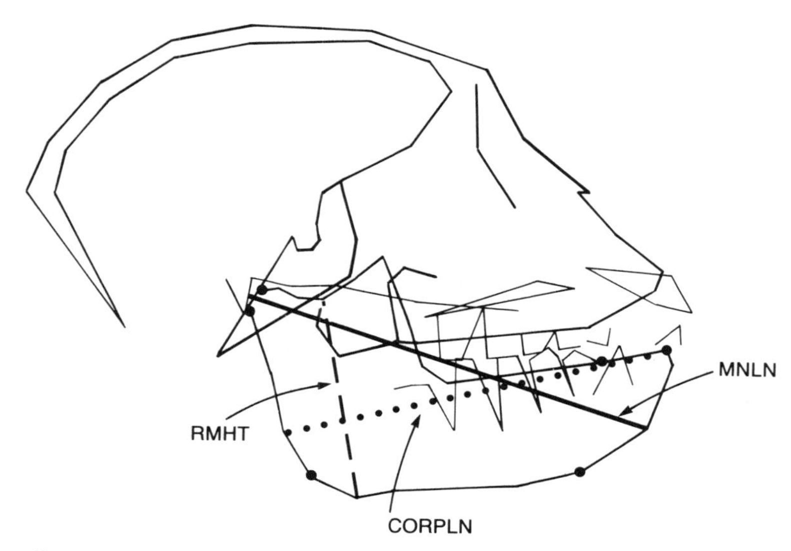

FIGURE 2.8 Mandibular dimensions and landmarks where remodeling has been measured.

sition of *menton* (63) relative to the mandibular reference line (MRL).

XGO1 and YGO1. Horizontal and vertical changes at *gonion* (68) relative to the mandibular reference line (MRL).

XCO1 and YCO1. Horizontal and vertical changes at *condylion* (72) relative to the mandibular reference line (MRL).

XSCO1 and YSCO1. Horizontal and vertical changes at *superior condylion* (73) relative to the mandibular reference line (MRL).

XPCO1 and YPCO1. Horizontal and vertical changes at *posterior condylion* (71) relative to the mandibular reference line (MRL).

YMNALVR1. Vertical changes in *mandibular alveolar ridge* in the region of the canine (at point 82) relative to the mandibular reference line (MRL).

MNLN. *Mandibular length* measured obliquely from pogonion (62) to condylion (72).

RMHT. Mandibular *ramus height* measured from the most in-

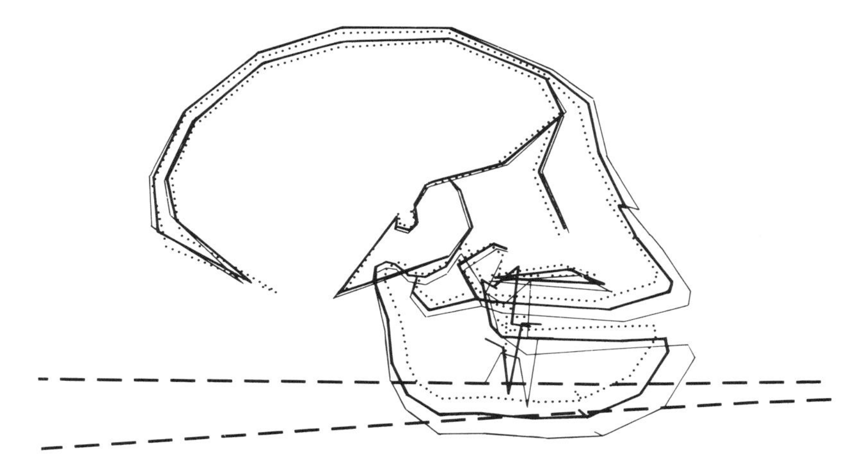

FIGURE 2.9 Angular displacement or rotation of the mandible relative to the cranial base (MRLCRL1) is shown in this composite of serial films superimposed on the cranial base implants. The broken line indicates the position of the MRL at two observations separated by approximately one year.

ferior point of the ramus (67) to the inferiormost point along the mandibular notch (76).

CORPLN. Mandibular *corpus length* as measured from the most anterior point along the mandibular alveolar ridge (83) and the most posterior point along the lower half of the posterior border of the ramus (69).

XMNDSPL1 and YMNDSPL1. Horizontal and vertical *displacement* of the anterior or symphyseal region of the *mandible* with regard to the cranial base. Operationally, the displacements of the symphyseal bone implant (169) relative to the cranial reference line.

MRLCRL1. Change in *angle* between the mandible (as represented by the mandibular reference line) and cranial base (as represented by the cranial reference line) (fig. 2.9).

MRLOCP1. Change in *angle* between mandibular body and mandibular functional occlusal plane. This occlusal plane was represented as in MXRLOPL1 (see above), except the mesial cusp of the *lower* first molar was used (point 106 in subadults, 153 in adults).

MRLMXRL1. Change in *angle* between mandibular and maxillary bodies.

Neurocranial Measurements
(Fig. 2.10)

FRNTH. Frontal bone thickness (distance between points 14 and 28).

PARTH. Parietal bone thickness (distance between points 16 and 26).

OCCTH. Occipital bone thickness (distance between points 19 and 23).

UPHT. Upper facial height (distance between points 29 and 38).

UPFDPTH. Upper facial depth (distance between points 1 and 30).

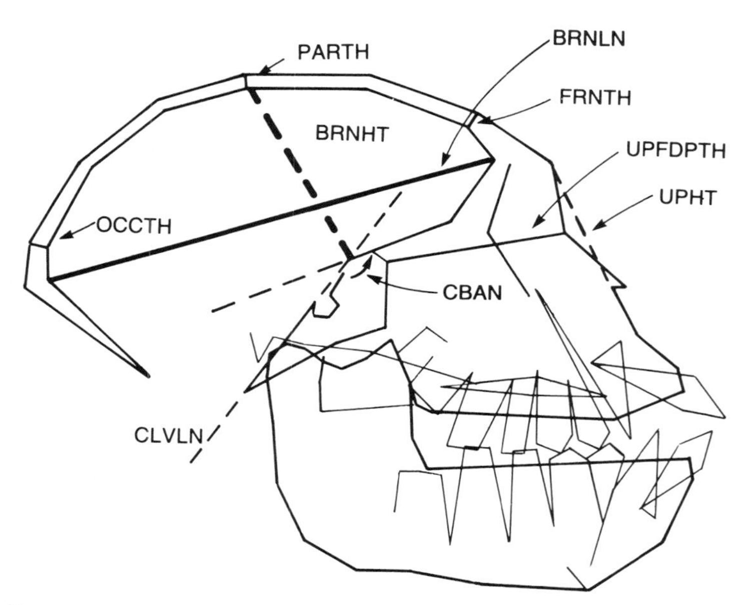

FIGURE 2.10 Neurocranial dimensions shown in a young adult male.

CLVLN. Clivus length (distance between points 4 and 5).
BRNLN. Brain length (distance between points 13 and 20).
BRNHT. Brain height (distance between points 10 and 16).
CBAN. Cranial base angle (angle between lines formed by points 12 and 10, and 5 and 4).

Dental Measurements
(Fig. 2.11)

Three types of measurements were taken for each of these teeth:

LDM2⟨measurement symbol⟩. Lower deciduous second molar.
UM1... Upper first permanent molar.
UM2... Upper second permanent molar.
UM3... Upper third permanent molar.
UC... Upper permanent canine.
LC... Lower permanent canine.
LM1... Lower first permanent molar.
LM2... Lower second permanent molar.
LM3... Lower third permanent molar.

Measurements

1. ⟨tooth symbol⟩ER. *Eruptive position.* The distance between the mesial cusp of the molar and the functional occlusal plane (see MXRLOCPL1 above). This is a measure of the *vertical position* of the tooth and completely erupted dentition.
2. ...AN. *Angle* of the distal root relative to the functional occlusal plane. This measure provides information on the *orientation* of the molar relative to the rest of the completely erupted dentition. This is of interest as erupting teeth move from an oblique to a parallel relationship as they move into the tooth row. (Not taken for canines.)
3. ...RTLN. *Root length.* Distance between distal cusp and

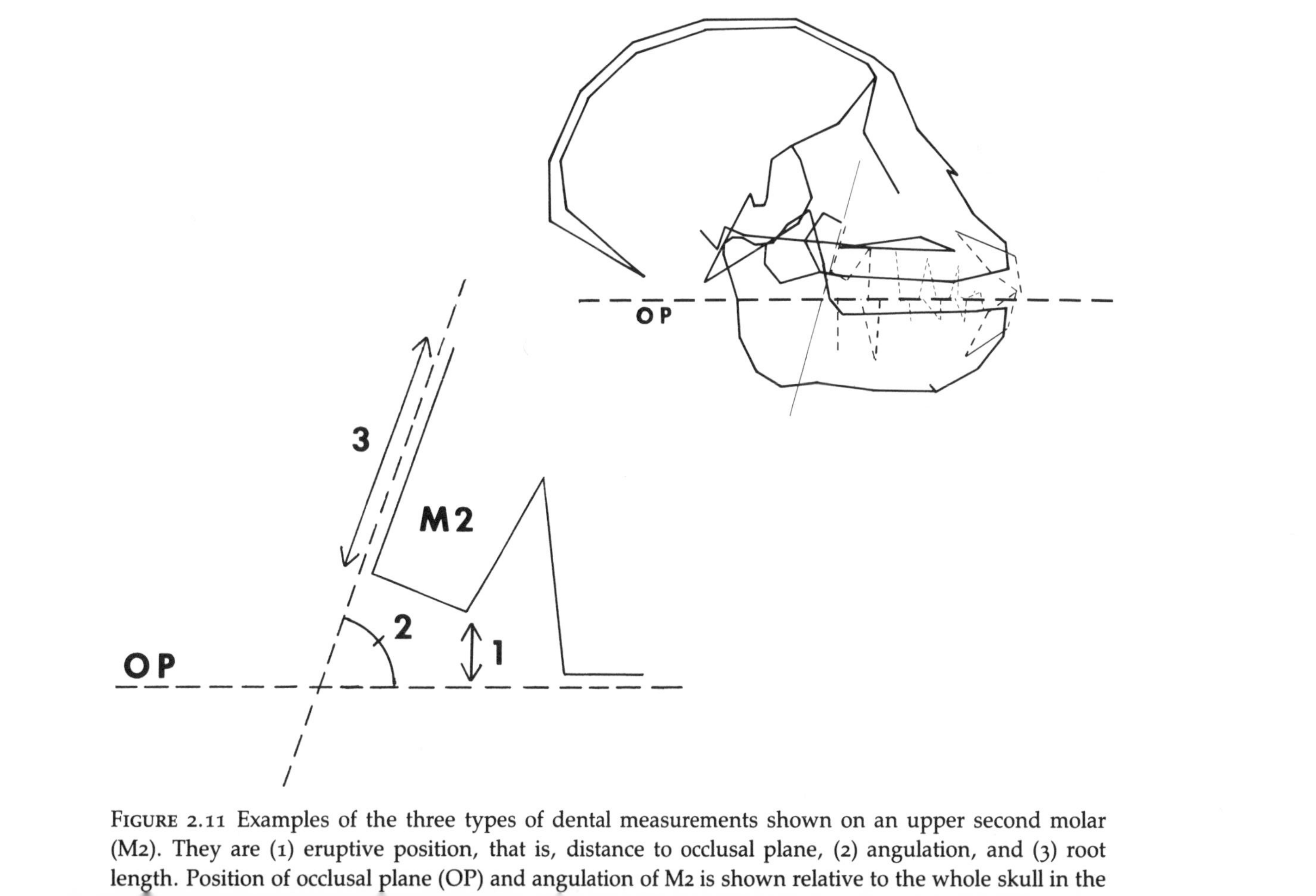

FIGURE 2.11 Examples of the three types of dental measurements shown on an upper second molar (M2). They are (1) eruptive position, that is, distance to occlusal plane, (2) angulation, and (3) root length. Position of occlusal plane (OP) and angulation of M2 is shown relative to the whole skull in the

distal root. This measurement is for comparative purposes only; it is not strictly a measurement of root length or maximum tooth length. (Not taken for the lower deciduous second molar, LDM2.)

Reference Line Construction

Three reference lines were constructed for superimposing serial films on bone implants in a standardized way, making computerized analysis possible. The *Cranial Reference Line* (CRL) was used for superimposing on the cranial base implants (specifically those in the body of the sphenoid bone). The *Maxillary Reference Line* (MXRL) was used for superimposing on the maxillary body bone implants. The *Mandibular Reference Line* (MRL) is used for superimposing on the mandibular symphysis and corpus implants. Each of these lines was defined on an early film of a series (ideally the first), and transferred to later films by means of the constructed points 174, 175, and 176. Each of these lines was defined as parallel to the functional occlusal plane (cf. Krogman and Sassouni 1957) on this early film. The three posteriorly located points that are involved in the reference lines were constructed as follows:

POINT 176. On the first film, the functional occlusal plane was identified. A bone implant located in the body of the sphenoid that was stable—that is, constant in orientation and position relative to other sphenoid implants and anatomic structures of that bone over a series of films—was identified (point 167). A line parallel to the original occlusal plane was drawn on the tracing of the first film passing through the sphenoid implant and extending posteriorly through the back of the cranium. A point was drawn on this film on the line near the ectocranial outline (point 176). The exact placement of this point on the first film was arbitrary; the only considerations were (1) that it be placed along the CRL on a radiopaque part of the film so that it shows through to subsequent films when they are overlayed, and (2) that it was at as great a distance from point 167 as consideration (1) will allow. This latter procedure served to minimize rotational error during the superimposition process (when done either by hand or computer). A fine-tipped felt marker capable of marking on a radiograph without smearing was used for locating these constructed points. The next film of the series was then superimposed on the sphenoid body implant (point 167) so as to produce a maximal correspondence between the anatomic details of the sphenoid as well as other sphenoid implants. After this was done, effort was made to maximize the correspondence of details in adjacent regions of the middle cranial base (e.g., ethmoid air cells, curvature of the greater wings of the sphenoid) while maintaining the correspondence between the more centrally located features of the body of the sphenoid bone, particularly the implants. Lastly, the position of the implants located in the frontal, temporal, and occipital bones was examined, when present. The final cranial superimposition was adjusted so as to minimize the displacements of these peripherally located implants, while preserving the alignment already established. The position of point 176 on the first film was then marked on the second film (as well as the tracing). Thus, points 167 and 176 serve to superimpose serial films on the central region of the sphenoid bone, orienting the two films so that the "horizontal" dimension for subsequent analysis is parallel to the original occlusal plane. Orientation could have been done with any plane that maintains the same superimposition. The original occlusal plane was chosen because it is a readily comprehensible horizontal and it is widely used in the craniofacial growth literature. The purpose of this procedure was to superimpose on the developmentally mature region of the central skull.

As the animals mature, the assignment of point 176 becomes less subjective. By the time the second permanent molars are formed in their crypts, much of cranial base and vault growth has been completed. At this stage it is often possible to simultaneously superimpose precisely on the implants in the sphenoid, frontal, and occipital bones. This is also true of the temporal implants somewhat later, presumably when lateral expansion of the temporal lobes of the brain is completed. Under these conditions, construction of the CRL is a simple and largely objective matter, requiring little interpretation of anatomic detail.

Conceptually the CRL can be viewed as the abscissa of a coordinate system based on the stable, unremodeled central portion of the cranial base. For example, when the translation of a maxillary body implant is described with regard to the CRL, its displacement/translation within this standardized cranial base coordinate system is being considered.

POINT 175. This point and the MXRL were constructed for superimposing on the maxilla in a fashion analogous to that used for point 176 and the CRL. A stable implant was located in the body of the maxilla, ideally in the region superior to the first molar (point 166). The stability of this implant was assessed by noting its precise position and orientation relative to the other maxillary body implants. The MXRL passes through this point and is parallel to the original occlusal plane. A point along the posterior end of this line was chosen using the same considerations used in the choice of point 176 described above. The MXRL can be viewed as the abscissa of a coordinate system based on the stable portion of the maxilla body. Remodeling of the maxilla at its various peripheral landmarks is described relative to this coordinate system.

POINT 174. This point and the MRL were constructed for superimposing on the unremodeled portion of the mandible in a fashion analogous to that used for the cranial base and maxilla. A stable implant was located and traced in the mandibular symphysis (point 169). The MRL passes through this point and is parallel to the original occlusal plane. A point along the posterior end of this line that is as close to the posterior margin of the ramus as possible was chosen as point 174. The same criteria used in the choice of points 176 and 175 were considered here. Similar to the CRL and MXRL, the MRL is the abscissa of a coordinate system based on the stable portion of the mandible.

STATISTICAL ANALYSES

Growth standards were computed in a variety of ways to accommodate varying sample sizes at various observations, as well as irregularities in the structure of the data set. Conventional univariate techniques and more appropriate custom multivariate techniques were applied to the Center for Human Growth and Development (CHGD) data set. Data were treated as (1) cross-sectional or (2) strictly longitudinal.

Cross-sectional Univariate Treatment

The mean, standard deviation, observed range, and 95% confidence interval were calculated for each of the maxillary and mandibular measurements at each half-year interval for each sex separately. These half-year intervals span the first 10 years of the life span of the rhesus monkey. Measures of kurtosis and skewness were also calculated to assess the univariate normality of these variables. This was done for each dimensional and angular variable as well as each of the change variables (deltas). Paired *t*-tests were performed to compare average change values to zero. Bonferroni corrections (Neter and Wasserman 1974; Alt 1982; Schneiderman and Carlson 1985) were used to minimize experimentwise error (i.e., overall error rate). For example, when fifteen *t*-tests were considered simultaneously, an attained significance level of 0.0033 was required for each test to maintain an overall significance level of 0.05 (0.05/15 = 0.0033). All of these statistics were calculated for the total (unstructured) data set using *MIDAS*, the interactive statistical package provided by the Statistical Research Laboratory at The University of Michigan (Fox and Guire 1976). The algorithms used by *MIDAS*

(1976) for computing these univariate statistics (excluding range) use ordinary least-squares methods.

While the descriptive statistics and tests described above make use of all of the observations collected, they are poor estimators of the true population parameters (cf. Tanner 1962). Ordinary least-squares estimations methods, when applied to longitudinal data, have been shown to provide estimates that are inferior to those generated by maximum likelihood methods (Elston and Grizzle 1962; Hoel 1964; Potthoff and Roy 1964; Rao 1966). The purpose of presenting these univariate measures is to provide summations of the available data where the more powerful multivariate methods cannot be applied because of small subsample sizes.

The means, standard deviations, and 95% confidence intervals provide a rough approximation of how the variables behave as a function of age. The nonsimultaneous (univariate) confidence intervals reveal time periods in which there is obvious overlap or divergence between the sexes but they are too imprecise to make definitive statements. In general, these statistics are *suggestive* of mean trends and of the variability associated with them in those parts of the data set for which the more rigorous methods (below) cannot be applied.

Statistical Analysis for Age Estimation

The univariate statistics described above have also been applied to the Wisconsin Regional Primate Center (WRPC) data set, which is strictly cross-sectional in nature (see chapter 3). Ordinary polynomial regressions of chronological age on each of the morphological variables individually were performed. For all of the variables, the coefficients of degree greater than three (i.e., higher than the cubic term) were insignificantly different from zero, so that equations of higher order were not used.

Stepwise regression (forward algorithm) was used to assemble the best sets of morphological variables for estimating chronological age. The nominal significance level for inclusion of a variable in an equation was set at 0.05. Variables that are highly intercorrelated and therefore redundant in the information they provide are excluded from equations generated by this method.

These analyses were done for each sex treated separately as well as together. Separate analyses were done for animals grouped according to gross maturational status and sex. Maturational status is defined dentally where "immature" refers to those without the full permanent dentition, "adult" to those with a dentition that is fully erupted into occlusion (except the canines). These assessments of "dental maturity" can be readily made from radiographs, allowing one to choose equations that may permit more accurate estimations than ones based on the whole series for that sex.

While the R^2 values and standard error (SE) values for the regression equations provide the best summaries of the accuracy of these equations in predicting chronological age *for members of the population upon which they are based,* they do not provide information on the utility of these equations in estimating age in other rhesus monkey populations. For the sake of argument (and only in this context), the two rhesus monkey samples used in this project were treated as representing separate populations, and the equations based on one were used to predict age in the other. The best equation for the appropriate age-sex category was applied to each of the thirty-five monkeys from the CHGD sample. Actual and estimated ages are compared for the eleven CHGD animals whose birth dates were known.

Correlation Analysis

Simple correlation analysis was performed on the full array of maxillary and mandibular *change* variables as an exploratory procedure. By examining the pattern and intensity of the Pearson product-moment correlation coefficients between maxillary and mandibular variables, it is possible to gain insight into the *coordination* of growth within the craniofacial skeleton (Solow 1966). However, unlike correlations within cross-sectional data sets such as those reported by Solow (1966) that mainly provide information regarding *proportions* or allometry, correlations between measurements of *change* extracted from longitudinal data provide information on the temporal coordination (or coincidence) of growth rates in various regions. Since many of the correlation coefficients calculated were significant at the 0.05 level, only the strongest correlations, attaining significance at better than the 0.001 level are considered in this study.

Longitudinal Multivariate Treatment

The longitudinal models that have been applied to this data set consider a number of time points simultaneously and take the corresponding covariance structure of the data set into account. Because the serial observations on a single measurement are considered the components or variables, these models are multivariate. The degree of the polynomial equation that fit the data best was determined using Rao's (1959) and Hills's (1968) approaches. Hills's approach was also used to calculate velocities and accelerations of the growth curves. The construction of the growth curve and the describing and testing of its various attributes were performed following Rao (1959), Elston and Grizzle (1962), and Schneiderman and Kowalski (1985, 1989). The parameters that define these polynomial growth curves were calculated using maximum likelihood estimators. *MIDAS*, *SAS*, and *GAUSS* were used to perform these computations.

Separate analyses were done for sets of three to five consecutive half-year intervals. The total time span was divided up in this fashion because (1) high-order polynomial equations needed to describe curves for long periods spanning a number of maturational periods are difficult to interpret biologically, and (2) degrees of freedom are conserved by minimizing the number of time points relative to the number of subjects in order to minimize confidence intervals (Schneiderman and Kowalski 1985).

Restructuring the Data Set and Dealing with Missing Data

Several procedures were used to create subsets of complete data. Such matrices in which all observations are present for all T time points for all N individuals are required to apply the multivariate methods described below.

When more than one observation was available for an animal for a particular interval (i.e., a half-year grouping), the observations were averaged. In addition to structuring the data set, this procedure also tends to dilute the effect of tracing error. Thus, for the multivariate tests, the sample size (N) is the number of subjects. For the univariate descriptive statistics, sample size (N.Ob.) is the total number of *observations* for a particular half-year interval, which exceeds the number of subjects in some instances. As these univariate statistics play an ancillary role in this thesis, the time savings gained by using N.Ob. rather than N (given the original data structure and the limitations of *MIDAS*) outweighed the cost of inaccuracy in the resulting measures of variability.

Subsets spanning anywhere from 1.5 to 3 years, in which all subjects had all but one or two observations, were obtained. Choosing a method for filling the missing data gaps presented a dilemma. No method exists for filling these gaps that will not bias the results. If the mean for all subjects at a particular time point are substituted into the gaps for that time point, then the variability at that time point will be artificially lowered and the confidence intervals would be spuriously narrow. Alternatively, one can interpolate the intermediate value from the values from the films on either side of the gap within a series for a given animal. This technique will tend to bias the trend line toward linearity. It will also tend to make the accelerations closer to zero. This latter course was chosen as the lesser of two evils, since the former would defeat the purpose of this study of generating reliable estimates of variability.

Hills's Approach

Hills's (1968) brief paper describes a test for finding the degree of the lowest-order polynomial curve that best fits a longitudinal data set by using the divided differences between observations. From the original matrix of observations made on a single variable, matrices of differences of the first order (velocities), second order (accelerations), and third order were calculated. From these it was possible to calculate the mean velocity $\bar{v}$ and acceleration $\bar{a}$ for each time point, as well as for each individual (averaged over all time points). The vectors of velocities ($\mathbf{v}$) and accelerations ($\mathbf{a}$) are calculated for each of the three segments of the growth curves (see below). Velocity and acceleration are readily interpretable quantities and therefore are useful to have in addition to the growth-curve parameters calculated using Rao's (1959) approach (see the next section).

Hills provides a procedure for testing whether a particular specification is adequate. Hills reports that this simpler and more direct test is completely equivalent to that developed by Rao (1959); the equivalence of these two tests was verified with these data as well as with a standard published data set (Elston and Grizzle 1962). Hills's method, though more obscure than Rao's (1959), was used in addition because it was inexpensive to process on the mainframe computer and provides informative adjunct statistics.

With Hills's method one determines whether the linear equation, which has two parameters, sufficiently fits the data by testing whether the vector of second-order differences ($\mathbf{a}$) is equal to zero. Similarly, one tests whether the quadratic equation is adequate by testing whether the vector of third-order differences is equal to zero, and so forth. The test itself involves a transformation from Hotelling's T^2 test (the multivariate analog of the Student's *t*-test) to the (univariate) F-distribution. A complete description and implementation of Hills's method can be found in Schneiderman and Kowalski (1989).

A more detailed description of the implementation of Hills's approach and its multigroup extension is in preparation (Schneiderman and Kowalski, in prep.). The statistical tests employed in the present study are limited to the longitudinal analysis of a single sample, that is, testing hypotheses about a single population. A method for formally testing hypotheses concerning between and among population differences for longitudinal data is currently being developed as an extension of Hills's method and *MANOVA* techniques.

Rao's Approach

Rao's method of growth-curve analysis (1959) was implemented using *SAS* (Schneiderman and Kowalski 1985). The best unbiased maximum-likelihood estimates for the true population growth-curve parameters (tau) are calculated using this method. The estimated polynomial growth curve is that which best fits the data with the least number of parameters. This method involves the calculation of a class of simultaneous confidence intervals for the average growth curve (CI). Graphically these are represented as a *confidence band*. It is these confidence bands which account for the *intercorrelation among serial observations* within the data set that distinguishes this approach from those currently in use in the growth and development literature. Unlike the means and standard deviations generated by the ordinary least-squares methods, which have been widely and inappropriately used to describe longitudinal data, the present approach will not underestimate the variability within a population, nor spuriously indicate significant trends where there are none.

This method also involves the calculation of confidence intervals for each of the parameters that reflect the precision of these estimations. The computer program written in the matrix language of *SAS* (SAS 1982a, 1982b), which is presented and documented in Schneiderman and Kowalski (1985), performs the complex computations involved in this method.

FORMAT OF RESULTS

Results are presented for each craniofacial measurement individually at each half-year time interval for the first 10 years of life. Emphasis is placed on the results of the multivariate tests (Rao's and Hills's). Unless stated otherwise, values discussed in the text are average values for particular time points and their 95% confidence intervals calculated by using Rao's method. Univariate means and the results of paired *t*-tests are also presented under circumstances in which the multivariate findings are ambiguous. Average velocities and accelerations (actually decelerations in most cases) are those calculated using Hills's method. These data are presented in tabular form in Appendix C and in graphic form in figures 4.1–4.19.

For the multivariate treatment of the males, the entire male data set was subdivided into three subsets for which there were complete blocks of observations for a maximal number of animals (i.e., perfectly rectangular blocks of data). Thus the entire study period is treated as three segments, 1.5 to 3 years, 3 to 6 years, and 6 to 7.5 years. The sizes of these male samples are thirteen, twelve, and eight, respectively. Average growth curves and 95% confidence bands (simultaneous confidence intervals) were calculated for each of the segments separately. Additional growth curves for segments with different starting and finishing time points were also calculated when the trends provided by the standard segments were ambiguous or enigmatic.

In figures 4.1 through 4.19, the average values for the male growth curves are represented by solid squares. The cross-hatched regions around these curves are the 95% simultaneous confidence regions. The open squares represent the means for the male samples calculated by conventional methods. These are referred to as *univariate* means in the text. They are presented to provide an impression of what is occurring during the periods before and after, and bridging the periods considered using the more rigorous multivariate methods. In some cases these means aid in the interpretation of the multivariate results. Those univariate means discussed in the text are based on smaller samples of six or more, unless stated otherwise.

The absolute values of those measurements that are strictly anatomical dimensions (i.e., distances between two landmarks; e.g., *maxillary length*) are plotted directly against the estimations of chronological age in years in figures 4.7, 4.8, 4.17, and 4.18). These are *growth* or *distance curves*. The corresponding velocities of these curves of absolute size are presented in tabular form. The change variables are also plotted against estimated age in years. These *growth velocity curves*, represented in the remaining figures in chapter 4, portray the change in rate of growth. Changes are expressed in mm or degrees per *half-year*, which is the unit time period used in this study.

Due to the small and irregular female sample, it was not possible to extract large blocks of continuous data on an adequate number of individuals so that the wholesale application of the multivariate methods to these data would be justified. The univariate means for the female sample are represented as open circles in figures 4.1 through 4.19. The purpose of presenting these data graphically is to *suggest* how the females might differ from the males. The female data must be considered as preliminary and interpretations of sexual dimorphism as speculative until a larger sample is gathered, and two-sample multivariate tests are developed and applied.

3

AGE ESTIMATION

A means of estimating chronological age in the rhesus monkey was required for placing the craniofacial growth and development of this species in a more precise temporal framework than is currently available. The resolution of this problem depended on the prior development of the cephalometric methods and measurements outlined in chapter 2. Because of the unique position that this issue occupies vis-à-vis the other problems addressed in this work and because age estimation techniques were developed on a separate cross-sectional sample of rhesus monkeys, this topic is considered separately in this chapter.

Solutions to the age estimation problem in the past have focused on changes within the dentition, body weight, height, and various other measurements that become larger as some function of age. Since the present investigation deals exclusively with cephalometric data, age estimation procedures have been confined to measurements that can be extracted from digitized cephalograms or measured directly from cephalograms or tracings with a pair of calipers and protractor.

Various regression methods were used on cephalometric data derived from the sample of captive rhesus monkeys from the Wisconsin Regional Primate Center (WRPC) for which the birth dates are known. A series of equations, which could then be used to estimate chronological age for other captive rhesus monkeys for whom this information is unknown, was developed for each sex. These equations were applied to the longitudinal data set from the Center for Human Growth and Development (CHGD), for which the birth dates were known for fewer than one-third of the subjects. The general utility of these equations based on the WRPC sample was evaluated directly by comparing estimations they generated with the known chronological ages for the eleven CHGD animals for which this information was available.

In addition to the pragmatic goal of generating the most reliable equations from which age can be estimated, another purpose of this part of the study was to explore the *relative* utility of various types of morphological variables, as well as various *types* of regression models.

BACKGROUND

It has long been recognized that development of the occlusion, that is, the number and kind of teeth that have erupted, is a good indicator of maturity in primates. Hurme and van Wagenen's (1953, 1956, and 1961) classic studies document the timing of eruption of the deciduous and permanent teeth in a large colony of rhesus monkeys. These data have been invaluable in general assessments of dental maturity and have enabled investigators to assign animals to gross developmental categories (e.g., infant, juvenile, adolescent, and adult). Although this information is adequate for many investigations (e.g., McNamara and Graber 1975; McNamara, Riolo, and Enlow 1976), a reliable means of distinguishing *among* animals within these categories is required to examine *subtle, short-term* developmental changes in morphology.

Two major investigations have studied chronological age estimation in the rhesus monkey. Gaven and Hutchinson (1973) evaluated three methods of age estimation in an analysis of captive and free-ranging rhesus monkeys, including (1) the total number of erupted teeth; (2) body weight and its cube root; and (3) sitting height. A series of univariate and multivariate regressions using these variables was developed on one of the captive samples consisting of fifty-one monkeys, all less than 90 months old. The average maximum deviation for all their equations was about 5.5 months.

Cheverud (1981) investigated the relationship between dental eruption, epiphyseal fusion, and chronological age in the skeletal collection of the Cayo Santiago rhesus monkeys (N = 299). Teeth were scored as erupted into occlusion or unerupted. Epiphyses at a number of joints were scored as fused or unfused. Regressions were calculated for each sex for all of the animals less than 85 months old using these dichotomous variables. Cheverud found that regressions for dental eruption scores alone and epiphyseal fusion scores alone were quite good ($R^2 > 0.95$, and standard deviations of the residuals were between 4 and 6 months), and even better for the combined dental and epiphyseal scores ($R^2 > 0.98$ and standard deviations of the residuals between 3 and 5 months). For all of these equations the regressions were significant at better than the 0.001 level.

APPROACH

A series of skeletal measurements relating to brain and eye development were taken to represent neurocranial maturity. It was postulated that cranial dimensions and angles relating to neural maturity would be the best estimators of chronological age. Because of the primacy of the development of the nervous system in higher organisms, the growth of these tissues might be more buffered from environmental influences than other tissues. In other words, the development of the neurocranial complex might be more conservative phylogenetically and ontogenetically (i.e., they are not as routinely subjected to environmental/biomechanical forces as the teeth and jaws) and should follow a more "narrow pathway" than the other tissues of the head (cf. Enlow 1982; Carlson 1985). For this reason the neurocranium should develop less variably as a function of time, within a population.

Previous studies using the dentition as an indicator of maturity have relied upon qualitative/discrete data, such as the number of erupted teeth or scores regarding the presence or absence of particular teeth. When available, it would seem that *continuous* data regarding dental position would be more informative. From digitized cephalometric data, one can encode a continuous range of states for the various teeth, rather than limiting oneself to discrete categories. Also, continuous data are more likely to be normally distributed and parametric in nature and therefore ideally suited for analysis by the conventional parametric statistical models such as the various regression techniques. Visual analysis of radiographs and composite plots suggested that the distance of erupting teeth from the occlusal plane, their angulation relative to it, and root length as they develop may be informative in age estimation.

A series of craniofacial variables representing the dimensions of the jaws were taken. Because of the highly adaptive and plastic growth potential of the jaws, as demonstrated in numerous primate and rat studies, it was hypothesized that these measurements would be highly variable as a function of chronological age and, therefore, least useful in age estimation.

Since the growth pattern of many dimensions of the craniofacial complex is curvilinear over periods spanning a number of years, it was anticipated that polynomial regression equations would fit these growth data better than simple linear regressions. Polynomial equations were calculated for all of the morphological variables in this study.

It was also hypothesized that a number of variables treated together should be more powerful in predicting age than any one variable alone. Vectors of variables may contain unique information that is not strictly additive. Therefore, chronological age was regressed upon sets of variables that included neurocranial, dentitional, and facial measures together as well as sets of variables representing a single system. With stepwise multiple regression it was possible to find the best equations from a great variety of possibilities systematically. The specific statistical methods involved in this portion of the study were detailed in chapter 2.

FINDINGS

A considerable number of variables, individually in polynomial equations (or linear equations in two cases) or in various combinations in multiple regressions, were found to be suitable for the prediction of chronological age (Appendix B). The equations are presented and discussed below according to gross maturational status and sex. All of the regressions presented are highly significant ($p < 0.0005$). Those equations having the lowest standard error in months (SE) and the highest R^2 values were deemed to be the best equations. The best as well as the *type* of equation that was most informative for each category is presented below.

Females

For all of the females together, the multiple regression equation using *hard palate length* and *ramus height* was best for predicting chronological age ($R^2 = 0.813$, SE = 14.83). Polynomial (quadratic) regressions using the lengths of the maxilla, mandible, and hard palate as well as ramus height were similarly predictive (SEs between 15 and 17 months). With the exception of ramus height, all of these variables reflect the sagittal depth of the lower face (prognathism).

For immature females, the multiple regressions equations using *permanent second molar root length* and various other positional measures of this tooth (from either dental arcade) and of the *deciduous second molar* were highly predictive of age ($R^2 = 0.944$, SE = 1.70, and $R^2 = 0.896$, SE = 2.31). One of these dental variables in conjunction with *palate length* was also quite good (R^2 = 0.918, SE = 2.35). In addition, the cubic equation of *ramus height* was good (R^2 = 0.832, SE = 4.81).

The only equation based solely on adult females that was of some value is the cubic of the *eruptive position of the upper third molar* (R^2 = 0.663, SE = 15.27). In general, the equations based on the entire female sample were as good, if not better, for predicting the ages of older animals.

Males

Like the equations for all the females together, those that provided the best estimations for the total male sample were primarily those measures reflecting the sagittal depth of the jaws. The best of these was the quadratic equation of *mandibular length* (R^2 = 0.941, SE = 8.24). Of interest is the multiple regression of age on the vector including *upper facial height, upper facial depth, and cranial base angulation* (R^2 = 0.834, SE = 13.92). This vector combines information on the size of the eyes and related tissues as well as the basicranial flecture, which reflects the spatial relationship between the neurocranium and the splanchnocranium (Baer and Nanda 1977). Though not as useful as the sagittal dimensions of the jaws (see above), this vector was approximately as predictive as the quadratic equations of *ramus height* and *maxillary height*, both of which reflect the vertical height of the lower face.

Age was best estimated in the immature males from a vector that includes the *eruptive position of the upper second molar* and *upper facial depth* (R^2 = 0.815, SE = 4.95). The linear equation of the root length of this same tooth was also of some use (R^2 = 0.637, SE = 6.93).

As with the females, the equations based on the adult males alone were no better than those based on the entire male sample. The multiple regression based on various aspects of the third molar position was the best for this category (R^2 = 0.819, SE = 11.03).

COMPARISON OF ESTIMATED AND ACTUAL AGES

Estimated and actual chronological ages were compared for a second sample of monkeys (CHGD) that was independent in all regards from the sample on which the equations were based (WRPC). Age estimations, actual ages, and the differences are presented below for the eleven animals from the CHGD sample with known birth dates (table 3.1). The equations used for these calculations are listed in Appendix B and are coded accordingly. Estimates were made on infant and juvenile films within each series using the best equations for the corresponding age-sex category. Ages are reported in months.

Note that the differences between the actual and estimated ages tend to be smallest when the calculations are made on the earliest films within the series; the estimations using craniofacial dimensions measured on infant films tend to be the most accurate. These results illustrate the general utility of these equations for estimating age in laboratory rhesus monkeys. It must also be noted, however, that these equations will occasionally provide misleading results (e.g., 353P), and that the estimations must always be considered in light of all the available developmental information.

DISCUSSION AND CONCLUSIONS

From these results it is apparent that excellent estimations can be made for immature monkeys, particularly females, when equations based solely on young animals are used. These equations based primarily on dental information are accurate to within ± 2 months for females and 5 months for males or juveniles of unknown sex. Accuracy is diminished considerably in older animals. When using the equations based on all males, some equations using jaw measurements are reliable to ± 8 or 9 months. For older females, an accuracy of only ± 15 months can be attained using an equation based on jaw measurements.

TABLE 3.1. Comparisons of actual and estimated ages for the CHGD monkeys with known birthdates.

Animal No.	Sex	Equation	Age Actual	Age Estimated	Age Difference
323N	M	M1	4.10	1.44	2.70
		M1	18.50	21.16	2.66
		M2	18.50	13.20	5.30
331A	M	M1	7.33	9.68	2.36
		M1	23.33	38.36	15.03
		M2	23.33	24.38	1.05
332L	M	M1	7.67	10.04	2.37
		M1	19.60	22.70	3.10
		M2	19.60	11.73	7.87
362B	M	M1	14.00	14.86	0.86
		M1	20.10	28.41	8.30
		M2	20.10	16.59	3.51
368D	M	M1	1.75	–0.87	2.63
		M1	25.27	34.60	9.33
		M2	25.27	21.49	3.78
369E	M	M1	1.63	2.41	0.77
		M1	28.23	36.46	8.23
		M2	28.23	21.49	6.74
1213A	F	F5	8.07	10.32	2.25
		F4	19.40	24.55	5.15
361A	F	F5	14.37	13.56	0.81
353P	F	F5	1.37	14.29	12.92
		F4	27.00	40.00	13.00
367C	F	F5	21.00	22.58	1.58
		F4	21.00	20.96	0.04
327H	F	F5	2.37	7.82	5.45
		F2	30.50	25.57	4.93

NOTES: Ages are in months. See Appendix A for listings of equations used here.

There is no advantage in using equations based solely on older animals. It was typical for variability to increase with age for the craniofacial measurements taken on this sample. Therefore, it was only advantageous to truncate the sample when deriving equations for estimating ages of young animals; the converse process was not advantageous for the older animals.

In general, the equations using sagittal dimensions of the

jaws have the widest utility. More precise estimations can be attained for animals in the mixed dentition stage by using equations that reflect the position and root length of teeth, particularly of the deciduous and permanent second molars.

Contrary to expectations, the neurocranial variables were found to be less useful in the prediction of age than the other types of data. The dimensions of the brain attain their adult proportions very early in life and appear to be quite variable among individuals. Both of these factors may limit their usefulness as predictors of age. However, it must be cautioned that the present findings may reflect the particular choice of measurements rather than any intrinsic informational limitation of neurocranial growth itself. Perhaps more complex variables, for instance, measures that reflect the volumetric expansion of the brain, would have been more useful in this context than simple linear dimensions.

Gaven and Hutchinson (1973) expressed some trepidation in the predictive accuracy of the age estimation methods they reported, that is, a maximum average deviation of a little less than 6 months. In light of the present findings, their simple methods are reasonably good. Consideration of subtle aspects of tooth positions and length and cephalometric dimensions does not radically improve one's ability to estimate chronological age.

Cheverud's (1981) method, which uses epiphyseal fusion in conjunction with dental eruption data, appears to provide some of the best overall estimations of chronological age; the use of these data would be particularly advantageous in older animals. Using the equations based on epiphyseal closure alone also has an advantage in the estimation of age in the context of craniofacial growth studies as it involves a system that is spatially independent of the dentofacial complex and apparently under separate hormonal control (Cheverud 1981). This method does, however, have the disadvantage of requiring the full skeleton or radiographs of the joints. Also, the computations are more complex, requiring the full variance-covariance matrices for the regression coefficients. Furthermore, these results may be less applicable to rhesus monkeys raised under laboratory conditions.

Even though they are no better than the equations offered by the previous reports on this subject, the equations provided in this study can be conveniently applied to radiographs. The measurements can be made directly on the radiographs or tracings with a pair of calipers and protractor, and the estimations made using no more than the regression coefficients provided and a calculator. These same equations could also be applied to skeletal material. However, unless measurements taken on dry skulls are taken in such a way as to be comparable to the cephalometric measurements, which are all projected onto the midsagittal plane, there would be some loss of accuracy. Also, some caution must be taken when applying such methods based on captive laboratory monkeys to samples representing other populations.

Though they may be frustrating, the present results are not surprising, as there is little biological basis for expecting the growth and development of a complex, long-lived mammal to bear a tight relationship with the passage of time. Environmental variability, which will have a compounded effect in those species having long periods of maturation, is superimposed on genetic variability in growth patterns.

Multiple regression equations that utilize polynomial variables may permit some refinement of the methods presented, improving accuracy in the estimation of chronological age. Ultimately, the best scale for assessing maturational status might be completely independent of chronological age. Change in some multivariate index that reflects proportional or shape changes due to maturation may fulfill this goal.

4

RESULTS

MAXILLARY GROWTH AND REMODELING IN MALES

Maxillary Displacement

Evaluation of the horizontal component of maxillary displacement (XMXDSPL1, fig. 4.1A), indicates that the maxilla displaces anteriorly at a strong and relatively constant rate of about 2.4 mm per half-year from infancy to about 3.5 years of age. During the 3.5- to 6-year span there is a significant deceleration in anterior displacement; the average deceleration is 0.41 mm per half-year^2 and the vector of decelerations differs significantly from zero ($p = 0.0007$). By six years the rate of displacement has diminished to about 0.7 mm per half-year. The remainder of the velocity curve, for XMXDSPL1 between 6 and 7.5 years of age, does not significantly differ from zero, perhaps due to the considerable variability around zero during this age period. The average trend does, however, suggest deceleration. The average deceleration of 0.26 mm per half year occurring between 6 and 7 years demonstrates a univariate difference from zero ($p = 0.015$); this is corroborated by the multivariate test in which $p = 0.06$. After 7 years the maxilla has clearly ceased to displace forward.

The horizontal component of maxillary displacement is considerably more prominent than the vertical component (YMXDSPL1, fig. 4.1B) in all age categories. At most observations,

MAXILLARY DISPLACEMENT

A. Horizontal

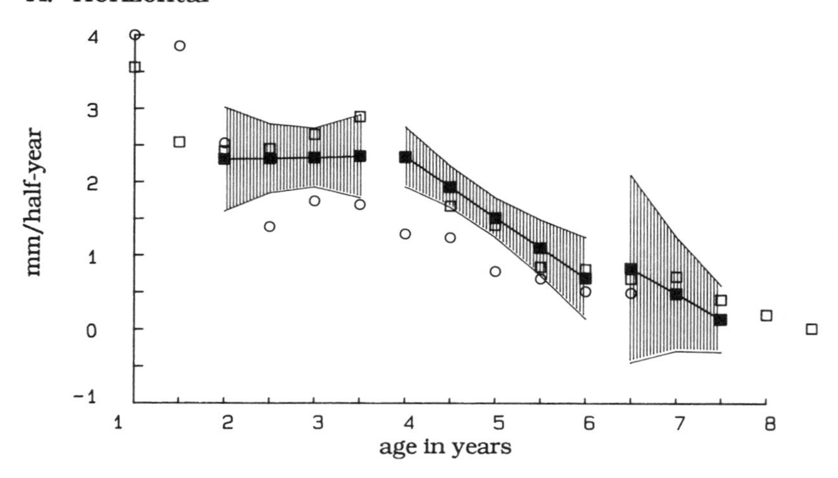

B. Vertical

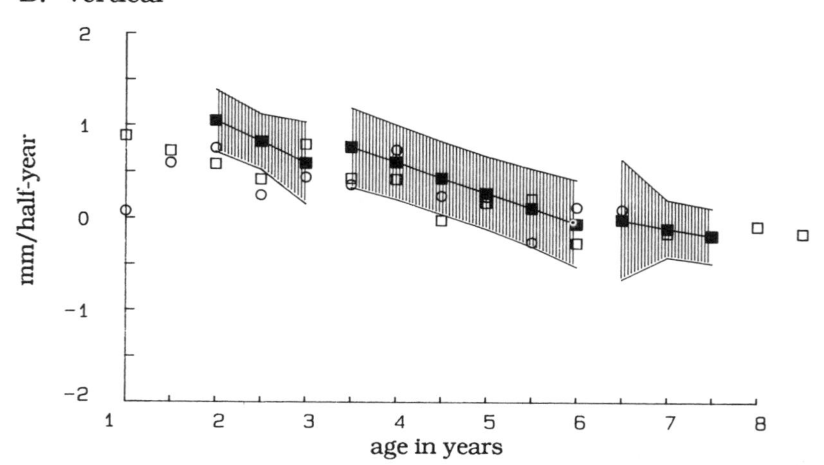

FIGURE 4.1 Velocity curves for the rate of *maxillary displacement* (X & Y-MXDSPL1), relative to the cranial base (CRL). Horizontal (A) and vertical (B) components of displacement. Velocities are expressed in mm per half-year.

the rate of inferior maxillary displacement is, at most, half of that in the anterior direction. The change in rate of vertical displacement is relatively constant; none of the decelerations for each of the time points differs significantly from zero over the entire curve. However, the vector of decelerations for the 3- to 6-year period is significantly different from zero ($p = 0.0074$), and the average deceleration for the earlier period from 1.5 to 3.5 years is also significant ($p = 0.0200$). The most striking contrast between the two components of maxillary repositioning is that the vertical activity ceases approximately 2 years before its horizontal counterpart. While there continues to be much variability in the 6- to 7-year period in horizontal displacement, the confidence band is narrowly distributed around zero in the comparable period in the vertical dimension. Though highly variable, the average direction of maxillary displacement (MXDSPLD) is 10 degrees over the course of the entire study period. Over the first 2.5 years of life it averages 20 degrees, over the remainder of the study period, 7 degrees.

Maxillary Rotation

The velocity curve for maxillary rotation (MXRLCRL1, fig. 4.2) illustrates that significant counterclockwise rotation occurs between 1.5 and 6 years of age. The male univariate means for the first 1.5 years of life indicate that the orientation of the maxilla to the cranial base is relatively constant during infancy (i.e., no significant rotation is detected, though the *direction* of the trend line anticipates the counterclockwise rotation that occurs over the next few years). During the 1.5- to 3.5-year span it rotates an average of 1.92 degrees. There is, however, a great deal of variability associated with this part of the curve (e.g., confidence bands are as great as ± 1.5 degrees at 2 years), making interpretation more difficult. Despite this variability, consideration of the growth curves generated by Rao's method in conjunction with the univariate means suggests a trend toward the gradual increase in the rate of counterclockwise rotation, with

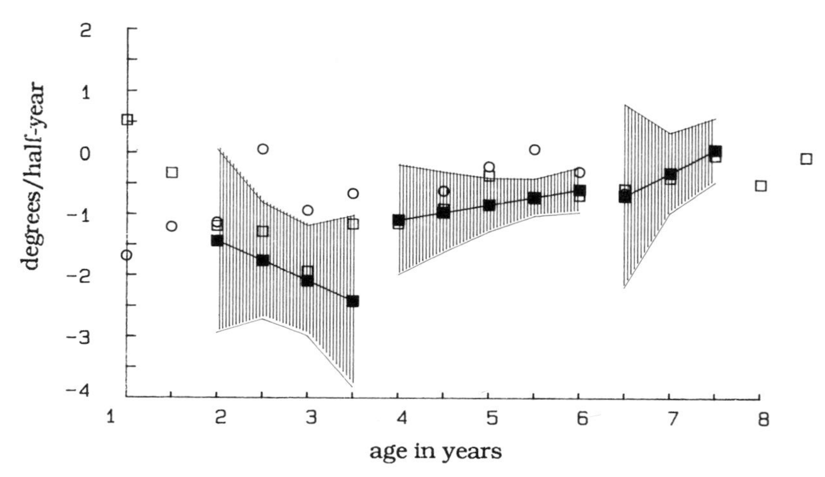

FIGURE 4.2 Velocity curve for the rate of *maxillary rotation* (MXRLCRL1) relative to the cranial base (CRL). The rate of angular rotation is expressed in degrees per half-year.

maximal activity occurring between 2.5 and 3.5 years, followed by a gradual decrease in rate as well as variability. After 6 years there is no significant evidence of change in the angular relationship of the maxilla to the cranial base.

GROWTH AT THE MAXILLARY TUBEROSITY

Changes at two points along the maxillary tuberosity (figs. 4.3, 4.4) were used to assess remodeling in this region. The maxillopalatine junction point (MXPJ) located at the inferior part of the tuberosity was found to be displaced posteroinferiorly relative to the maxillary body, indicating that bone apposition was occurring there. The velocity curve representing the horizontal component of growth at the maxillopalatine junction point (MXPJ, fig. 4.3A) very closely parallels that for the horizontal displacement of the maxilla in both shape and magnitude (see above). In contrast, however, the vector of decelerations for the 4- to 6-year segment of the curve does not significantly differ from zero. This is perhaps attributable to the greater variability in the horizontal displacement of the maxillopalatine junction point (XMXPJ1) than that observed for the horizontal displacement of the maxilla (XMXDSPL1).

The velocity curve for the vertically directed component of this growth at the maxillopalatine junction (YMXPJ, fig. 4.3B) demonstrates a confidence band that includes zero throughout the entire ontogeny. Vertical growth at this landmark is marginal; none of the vectors of velocities or accelerations for each of the three segments of the entire curve differs significantly from zero. On the other hand, the average velocities for the first and second segments of the total curve of 0.49 and 0.32 mm per half-year each demonstrate univariate differences from zero at the 0.02 and 0.06 levels, respectively. Furthermore, more than 75% of the confidence band between 2 and 3.5 years of age is above zero, suggesting that there is real vertical growth in the majority of male rhesus monkeys during this period. There is also a suggestion of a growth peak at 3 to 3.5 years of age; the individual velocity during this period is univariately different from zero at the 0.05 level. By 5.5 years, any vertical growth at this landmark has clearly ceased. The overwhelming predominance of horizontal over vertical growth at MXPJ results in an average vector of growth for all observations that is about 3 degrees above the original occlusal plane, that is, nearly purely horizontal.

Posteriorly directed growth at the pterygomaxillary fissure, inferior point (XPTMXFI1, fig. 4.4A) is strong during the first 3.5 years of life, averaging 1.73 mm per half-year. Decelerations over this time span are not significant. At 3.5 to 4 years of age there is graphic evidence of a growth peak of about 2 mm per half-year. From this point on there is clear-cut deceleration; the vector of decelerations corresponding to this segment of the curve is highly different from zero ($p = 0.0009$). The average deceleration is 0.35 mm per half-year^2. From 5.5 to 7.5 years the evidence for significant growth is equivocal; the vector of velocities does not differ from zero, but more than 75% of the confi-

MAXILLOPALATINE JUNCTION

A. Horizontal Growth

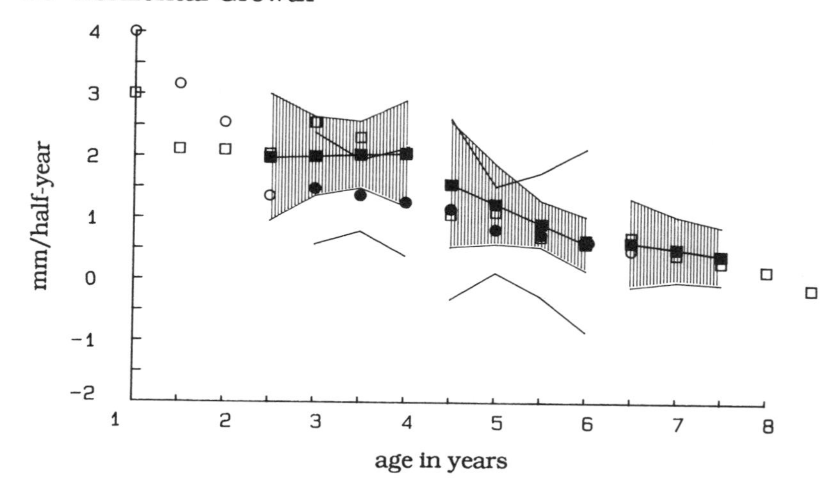

B. Vertical Growth

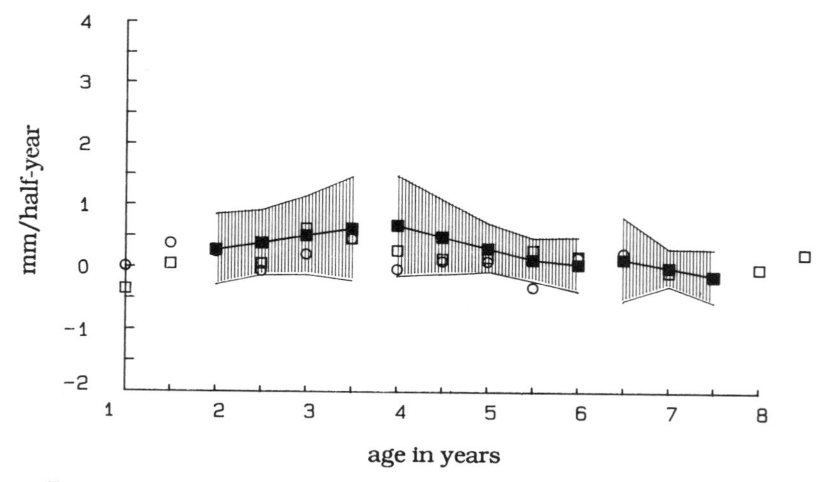

FIGURE 4.3 Rate of growth at the *maxillopalatine junction point* (X & Y-MXPJ1) in mm per half-year. Specifically, these measurements are (A), the horizontal displacement of the landmark relative to the body of the maxilla (MXRL), and (B), the vertical component of the displacement of this landmark relative to the MXRL. The solid circles and blank confidence bands in (A) represent the female values calculated using Rao's method.

PTERYGOMAXILLARY FISSURE

A. Horizontal Growth

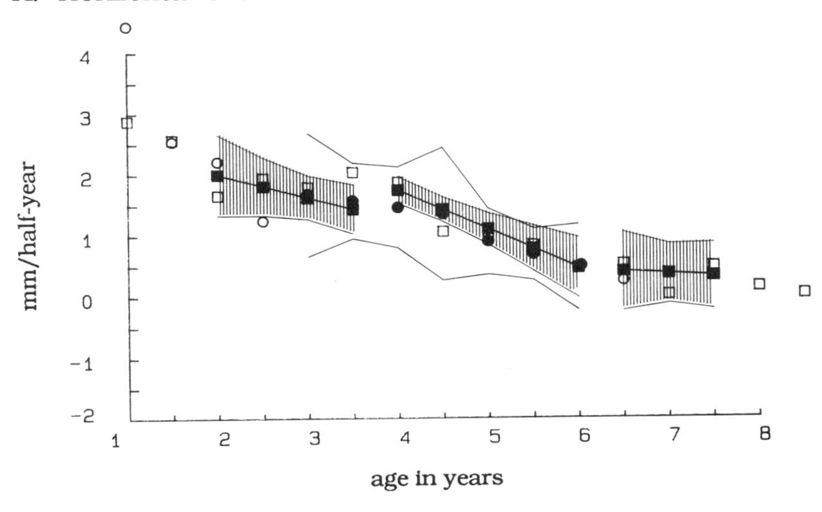

B. Vertical Growth

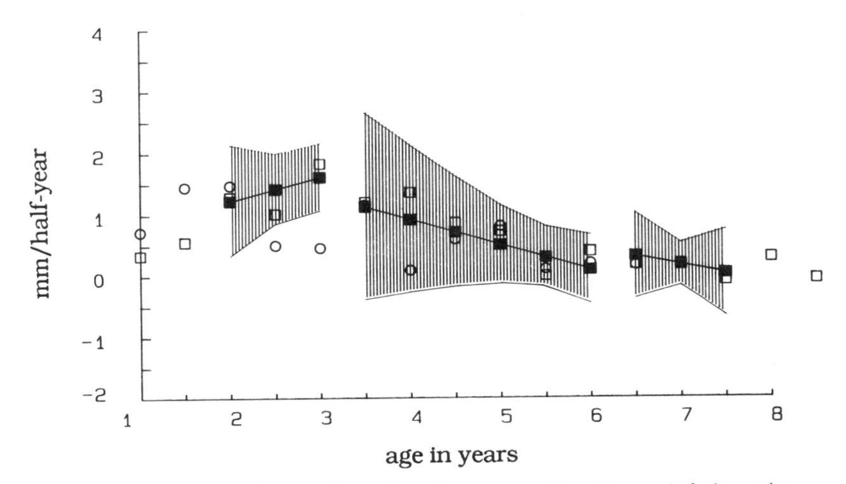

FIGURE 4.4 Rate of growth at the *pterygomaxillary fissure inferior point* (X & Y-PTMXFI1) in mm per half-year. (A) Horizontal displacement, and (B) vertical displacement of the landmark relative to the body of the maxilla (MXRL). See figure 4.3 caption for symbols.

dence band is above zero and univariate tests show the 6.5- and 7.5-year velocities to differ from zero at better than the 0.05 level.

In contrast to the vertical growth at the maxillopalatine junction point (MXPJ), the pterygomaxillary fissure inferior point (YPTMXFI1, fig. 4.4B) displaces inferiorly at rates that significantly differ from zero until about 5 years of age. The shape of this curve resembles that of the vertical growth at the maxillopalatine junction point, although it is of greater magnitude. Vertical growth at the pterygomaxillary point also peaks at about 3 years. Thus, vertically directed growth at the superior aspect of the maxillary tuberosity would appear to contribute significantly to gains in maxillary height and to downward displacement of the entire maxillary complex. The behavior of the pterygomaxillary point in the vertical dimension (YPTMXFI1) is highly variable over the 3.5- to 4.5-year period, as indicated by the broad confidence band.

The direction of growth (averaged over the whole study period) at the pterygomaxillary fissure, inferior point (PTMXFI) is about 22 degrees above the occlusal plane. The majority of the conventional 95% confidence intervals for the individual observations fall between zero and 45 degrees. Only at 3 and 6 years is the average direction at the high end of the range. Thus, horizontally directed growth predominates at the upper part of the maxillary tuberosity, but not to the extent that it does at the more inferior tuberosity landmark, the maxillopalatine junction point (MXPJ).

Remodeling at Supradentale

Anteriorly directed growth at supradentale (XSD1, fig. 4.5A) is significant throughout the 7.5-year ontogeny. The vectors of velocities for the three segments of the curve differ from zero at the better than or equal to the 0.05 level. Deceleration in growth at this point is quite gradual over the entire curve. Average velocities for the three segments drop from about 1 mm to 0.73 mm to 0.31 mm per half-year for each of the segments, but

SUPRADENTALE

A. Horizontal Growth

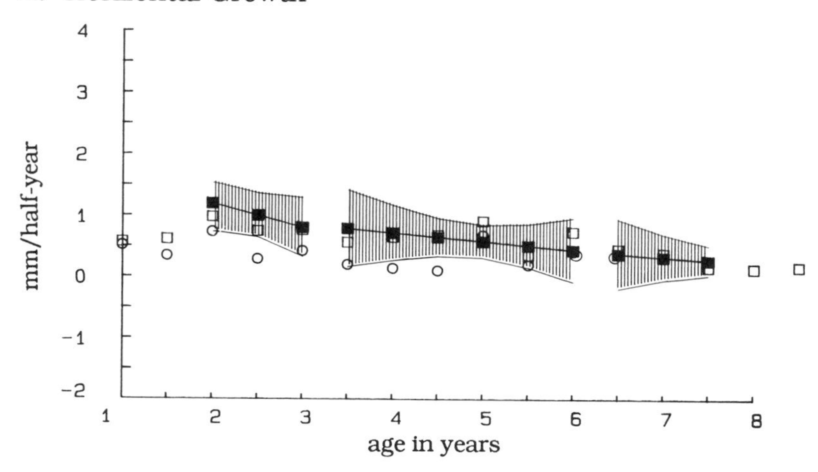

B. Vertical Growth

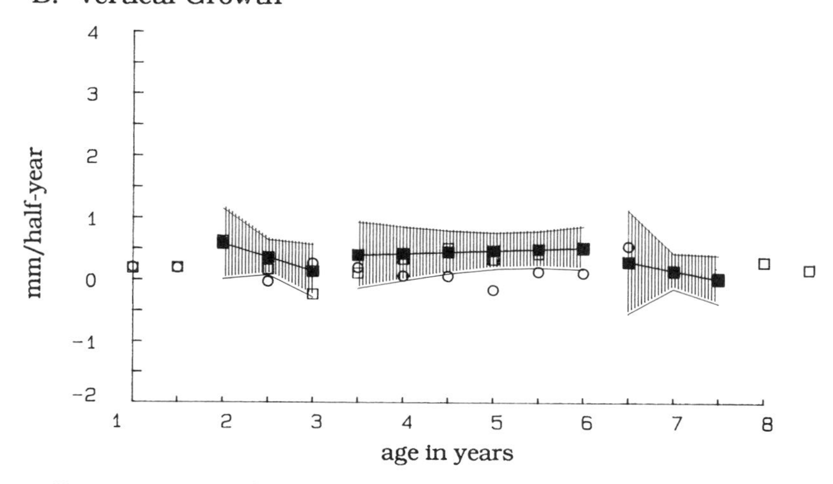

FIGURE 4.5 Rate of growth at *supradentale* (X & Y-SD1) in mm per half-year. (A) Horizontal and (B) vertical displacement of this landmark are relative to the body of the maxilla (MXRL).

only the vector of decelerations for the 1.5- to 3-year segment of the curve approaches significance ($p = 0.06$). Examination of the individual velocities and the confidence band over the 6- to 7.5-year period suggests that anterior displacement has ceased in many males by 6.5 years, and is effectively zero in all by 7.5 years. Some low level but univariately significant activity at around 7 years may account for the multivariate significance of this last segment of the curve.

Changes in the vertical position of supradentale (YSD1, fig. 4.5B) are quite modest throughout the study period. Initially, at 2 years it displaces inferiorly at about 0.58 mm per half-year. By 3 years this rate of displacement is effectively zero. Over the 3-to 6-year span, vertical growth here is quite constant, with an average velocity of 0.44 mm per half-year. From 6 to 7.5 years vertical change here is nil. In combination, these orthogonal measures of the displacement of supradentale (XSD1 and YSD1) provide a clear-cut picture of anteroinferior repositioning (with the anterior component predominating) of this landmark.

Maxillary Alveolar Ridge

The inferior repositioning of the maxillary alveolar ridge at the canine (YMXALVR1, fig. 4.6) occurs at a moderately strong and relatively constant velocity of about 0.87 mm per half-year from 1.5 to 3.5 years. After this phase, increments drop to a constant average rate of 0.44 mm per half-year until about 6 years. Behavior at this landmark between 4 and 7.5 years of age is remarkably similar to the vertical growth at supradentale (YSD1) occurring over the same time span. After 6 years of age there is no evidence of vertical growth at either point.

Maxillary Length

Total maxillary length (MXLN, fig. 4.7) increases at a very constant average rate of about 3 mm per half-year over the first 3.5 years of life. There is remarkably little variation around the seg-

MAXILLARY ALVEOLAR RIDGE

Vertical Growth

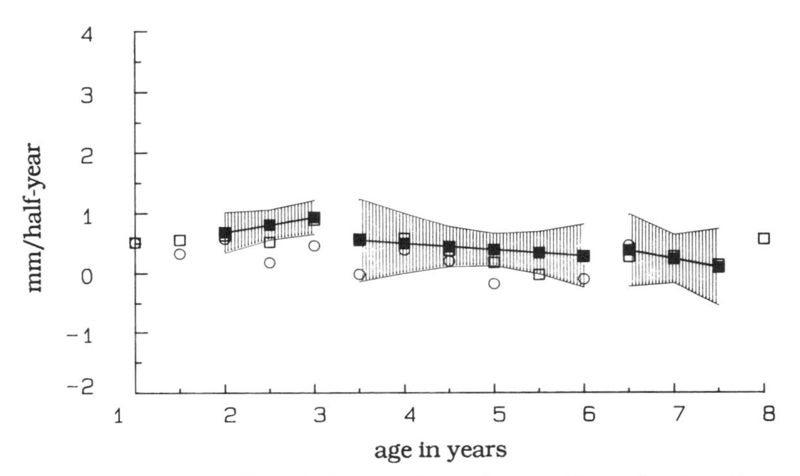

FIGURE 4.6 Rate of vertical growth at the *maxillary alveolar ridge* (YMXALVR1), that is, the vertical displacement of a point immediately posterior to the canine tooth relative to the body of the maxilla (MXRL).

ment of the growth curve between 1.5 and 3.5 years (ranging from 32.16 ± 2.14 to 44.04 ± 1.29 mm). The next segment of the curve for the period between 3.5 to 6 years of age initially continues along the same trajectory, but later undergoes significant deceleration. The average velocity for this segment of the growth curve is 2.27 mm per half-year and the average deceleration is 0.43 mm per half-year^2. The vector of decelerations is significantly different from zero ($p = 0.01$). A quadratic equation is necessary to fit the data adequately when observations from 3 to 6 years are considered together. There is a leveling in the rate of increase in maxillary length after 6 years of age. For the last segment the average velocity is 0.76 mm per half-year. The vector of velocities differ from zero at $p = 0.05$. Each of the individual velocities from 6 to 7.5 years demonstrate univariate

MAXILLARY LENGTH

A. Total Length

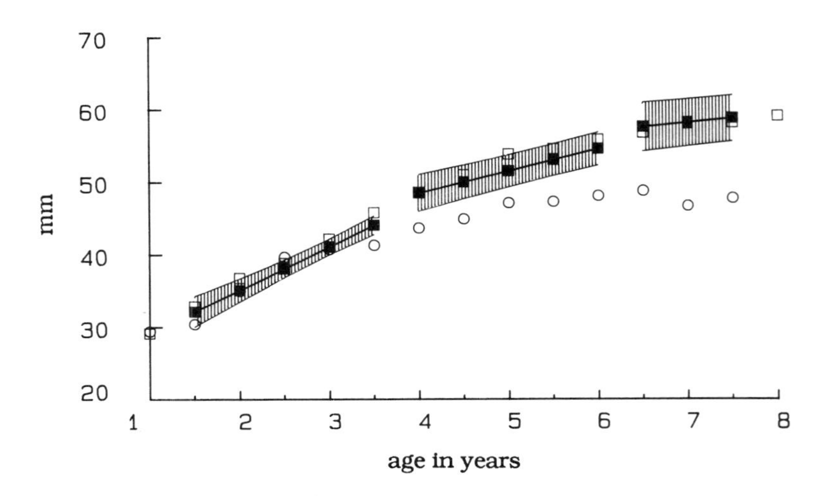

B. Change in Length

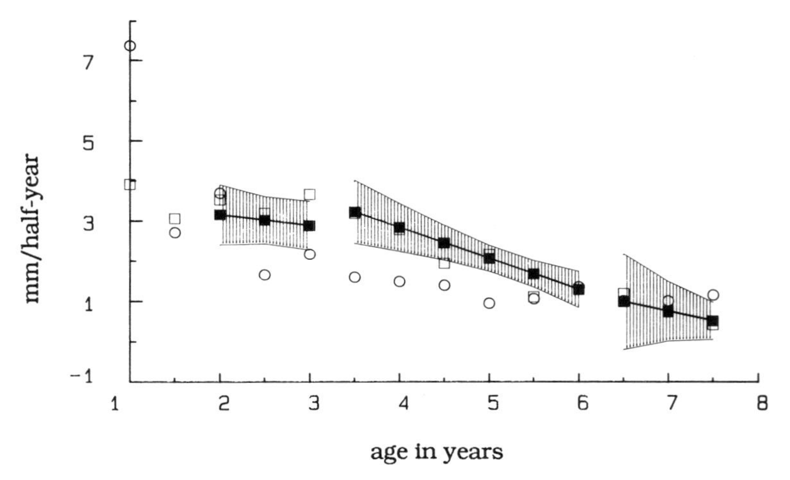

FIGURE 4.7 (A) Growth curve for the total length of *maxilla* (MXLN) as measured from the most anterior point on the maxilla to the maxillo-palatine junction point on the maxillary tuberosity. (B) Velocity curve for growth in maxillary length (MXLN1). Increments in maxillary length are expressed in mm per half-year.

significance at better than the 0.03 level. Decelerations for this part of the curve are insignificant. The data provide firm evidence that the maxilla of males is still lengthening at 6 years. The univariate means based on the scanty samples of older male animals suggests that total maxillary length plateaus at about 7 years of age at an average of 58.2 ± 3.27 mm.

Maxillary Height

In contrast to maxillary length, the growth curve for posterior maxillary height (MXHT, fig. 4.8) does not demonstrate any significant accelerations for any of its segments; it is fit aptly by linear equations throughout. Examination of the entire curve, which has a narrow confidence band up until about 6 years, reveals a gradual decrease in rate. Posterior maxillary height stabilizes at an average of 19.2 ± 3.76 mm at about 6 years; after this time there is no statistical evidence of growth.

Summary of Maxillary Growth and Remodeling

The maxilla displaces forward at a high and constant rate from infancy until about 4 years of age. Significant deposition of bone along the posterior aspect of the maxillary tuberosity (both at its most superior and inferior extents) accounts for the majority of this forward displacement of the maxilla. The small proportion of forward displacement not accounted for metrically by addition of bone at the tuberosity is perhaps attributable to radiographically undetectable growth at the vertically oriented sutures that are interposed between the maxilla and the sphenoid.

Over the next period, horizontal growth velocity and deceleration is virtually identical at the two maxillary tuberosity landmarks. However, it should be noted that the activity at the inferior landmark is considerably more variable. The significant deceleration at both of the landmarks resembles that in the horizontal displacement of the maxilla as a whole. Additional pro-

MAXILLARY HEIGHT

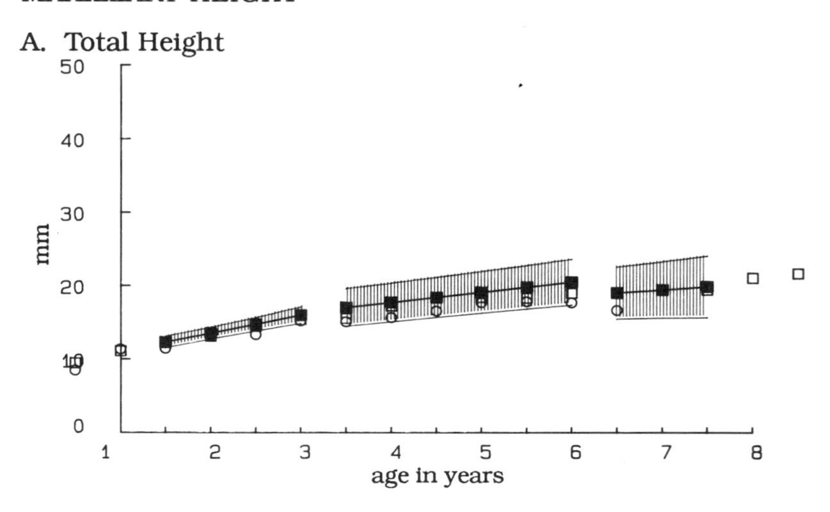

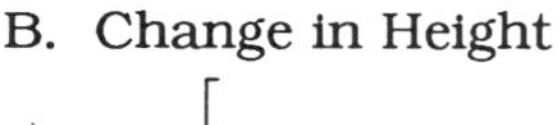

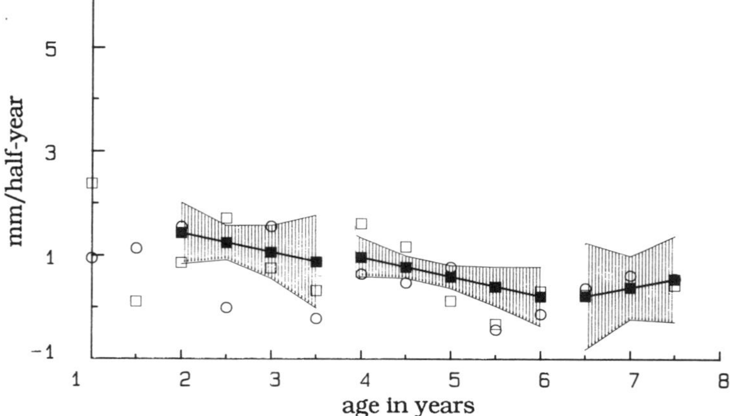

FIGURE 4.8 (A) Growth curve for total *maxillary height* (MXHT) as measured posteriorly from a point along the maxillary alveolar ridge that is distal to the last molar to the most superior point on the pterygopalatine fossa. (B) Velocity curve for growth in maxillary height (MXHT1). Increments are expressed in mm per half-year.

portions of this displacement must be accounted for by a source other than growth at the maxillary tuberosity since there is a disparity in average velocities over this time span. This disparity suggests that the sutures between the maxilla and the core of the facial skeleton may be relatively more active during this phase of ontogeny.

During the 6.5- to 7.5-year span, a cessation of anterior maxillary displacement corresponds to a concurrent cessation of significant horizontal growth at the tuberosity.

Vertical growth at the two maxillary tuberosity landmarks is considerably less pronounced than its horizontal counterpart. This corresponds well with the observation of markedly less displacement of the maxilla in the inferior direction. The maxilla displaces inferiorly relative to the cranial base at about one-third the rate of anterior displacement up until about 3.5 years, and at about one-sixth of the rate over the 3.5- to 6-year period. Addition of bone at the superior aspect of the maxillary tuberosity, as reflected by vertical growth at the pterygomaxillary fissure, inferior point (PTMXFI), would appear to play an important role in the overall inferior displacement of the maxilla, though vertical growth activity at this site is quite variable between 3 and 5 years of age (much more variable than the measure of vertical displacement of the maxilla itself, YMXDSPL1).

Average vertical increments at the two maxillary tuberosity sites sum approximately to the total increments in posterior maxillary height during the early phases of growth. Vertical growth peaks at the these sites at 3 or 4 years of age correspond temporally to the only velocity for the maxillary height velocity curve between 3.5 and 6 years that attains univariate statistical significance ($p = 0.003$). The relationship between vertical activity at these two tuberosity sites and increments in maxillary height is less clear-cut during the later phases of ontogeny.

Although overall lengthening of the maxillary complex is accounted for to a great extent by addition of bone along the posterior aspect of the maxillary tuberosity, a smaller but significant amount of primarily anteriorly directed deposition of bone at supradentale also makes a significant contribution (20% to 30%) to this overall lengthening.

The horizontal increments at the tuberosity and supradentale do not simply sum to the total increments in maxillary length. It is probable that growth at the premaxillary suture accounts for some of these increments, as indicated by the separation of premaxillary and maxillary body implants on the composite plots. An additional source for this disparity may be geometric artifact. The depository activity at these sites is predominantly but *not strictly* horizontal; there is a smaller but real vertical component to the growth at these sites that changes the resultant vectors of growth. The line that is used to represent maxillary length (between point 41, the most anterior point on the maxilla, just above supradentale, and point 45, the maxillopalatine junction point) changes its orientation relative to the unremodeled core of the maxillary body as these two landmarks relocate relative to it. Thus, *maxillary length* (MXLN) measures two slightly different dimensions at any two observations.

Composite plots of the maxilla indicate that the anterior displacement of supradentale is characteristic of the gradual apposition of bone that occurs along the entire facial aspect of the maxilla. Appositional activity here decreases very gradually throughout ontogeny (ceasing at about 7 years), though it is relatively more constant in rate than that occurring at the tuberosity. Significant inferior repositioning of supradentale occurs until 6 years of age and appears to occur in concert with the addition of bone at the alveolar ridge.

When all of the growth curves and univariate means for measurements taken at the maxillary tuberosity and alveolar ridge are considered together, there is an indication of a growth peak at 3 or 4 years of age. Also during this period, the greatest counterclockwise rotation of the maxilla occurs. As there is no evidence of statistically significant accelerations during this period, it is not appropriate to describe this as a growth spurt. There are, however, significant decelerations *after* this period in a number of these measures.

MANDIBULAR GROWTH AND REMODELING IN MALES

Mandibular Displacement

The repositioning of the mandible was determined by following the displacement of the implants in the symphyseal region relative to the cranial base. Thus, the following description is limited to the repositioning of the *anterior* end of the mandible. The horizontal component of this mandibular displacement is shown in figure 4.9A. The description of the rotation of the mandible (see below) provides information on the displacement of the gonial or posterior region of the mandible.

The average anterior mandibular displacement is quite significant and reasonably constant over the first 3.5 years of life (3.26 mm per half-year; $p < 0.0001$). There is no evidence of acceleration over this same period. Over the 3.5- to 6-year period this velocity curve is distinctly sigmoid in shape and is best fit by a cubic equation. In addition to the vector of velocities (first-order differences), and accelerations (second-order differences) being highly significant ($p < 0.0001$ and $p = 0.0015$, respectively), the vector of third-order differences is also highly significant ($p = 0.0020$). In more general terms, what this means is that after a precipitous drop in the velocity of anterior displacement after 4 years (deceleration at this time is 1.33 mm per half-year^2), there is an indication of a resurgence of anterior repositioning of the mandible between 5.5 and 6 years.

Between 6 and 7.5 years, the rate of anterior mandibular displacement averages 0.93 mm per half-year. The vector of the velocities over this period is significant at the $p = 0.057$ level, whereas univariately these velocities, as well as their average, are all significant at better than the 0.01 level. The slope of the average growth curve of this period, when extended into the univariate means available for 8 and 8.5 years, indicate that this anterior displacement is decreasing gradually to zero. Over the entire curve, especially between 4 and 5 years, there is a great deal of variability in anterior mandibular displacement

MANDIBULAR DISPLACEMENT

A. Horizontal

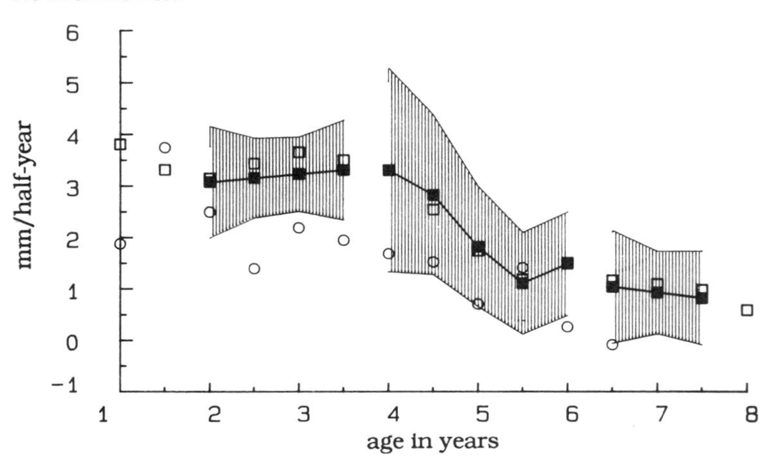

B. Vertical

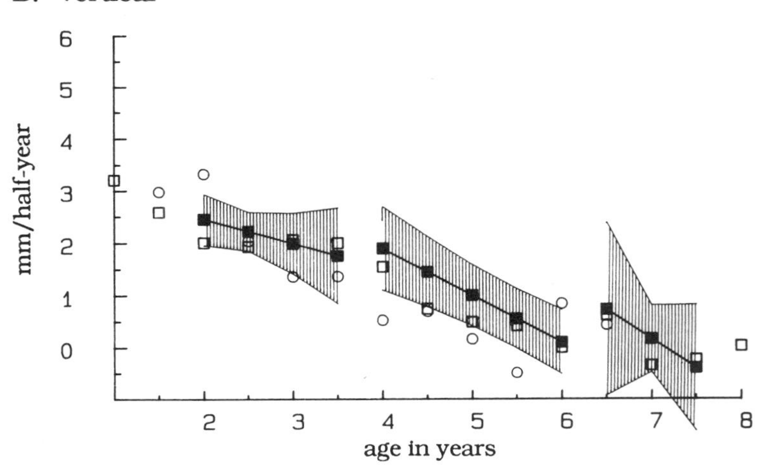

FIGURE 4.9 Velocity curve for the rate of *mandibular displacement* (X & Y-MNDSPL1) relative to the cranial base (CRL). (A) Horizontal and (B) vertical component of displacement. Velocities are expressed in mm per half-year.

(XMNDSPL1). Despite the very wide confidence bands, the trends described above are statistically significant.

The confidence band for the velocity curve for the vertical component of mandibular displacement (YMNDSPL1, fig. 4.9B) is considerably more narrow than that for its horizontal counterpart. Also the shape of the curve is much simpler. The first two segments of this curve are well fitted by linear equations. From infancy to 5.5 or 6 years the rate of inferior displacement decreases gradually. The vector of decelerations from 3 to 6 years is quite significant ($p = 0.0008$). The average deceleration for this period is 0.43 mm per half-year^2. From 5.5 or 6 years of age on, the velocity and acceleration of this variable is effectively zero.

Mandibular Rotation

The velocity curve for the rotation of the body of the mandible (MRLCRL1, fig. 4.10) indicates that it undergoes significant counterclockwise rotation relative to the cranial base from infancy through 6 years of age. From infancy to 3.5 years of age the

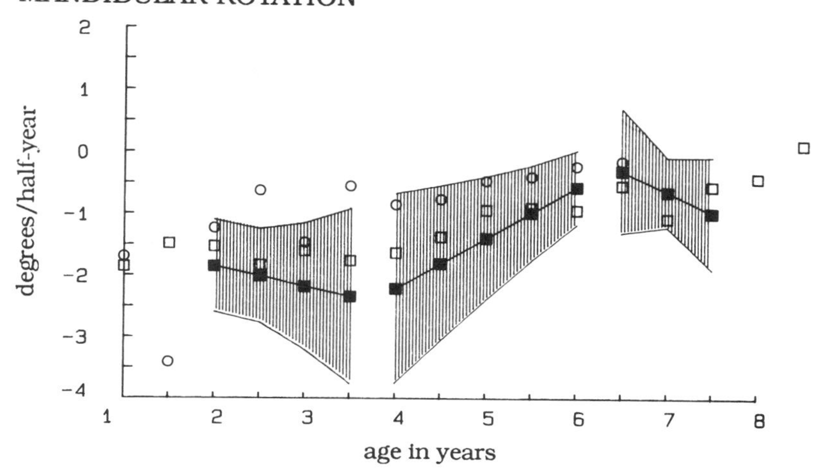

FIGURE 4.10 Velocity curve for the rate of *mandibular rotation* (MRLCRL1) relative to the cranial base (CRL). The rate of angular rotation is expressed in degrees per half-year.

average rotation is –2.21 degrees per half-year and is reasonably constant at this rate over this period, as indicated by the failure of the vector of accelerations to attain significance. There is a peak of rotational activity of about –2.3 degrees at 3.5 to 4 years of age. From this age until 6 or 6.5 years the rate of counterclockwise rotation decelerates significantly to about 0.57 degrees (the vector of decelerations is significant at the $p = 0.0032$ level). Except for brief but significant counterclockwise rotation of about 0.6 degrees per half-year at either 7 or 7.5 years, the angular position of the mandible appears relatively stable after 6.5 years.

Growth at Mandibular Condyle

Growth activity at the various mandibular landmarks is assessed relative to the mandibular reference line (MRL). Posteriorly directed growth of the condyle at condylion (XCO1, fig. 4.11A) is initially quite strong and decreases almost linearly throughout ontogeny. Over the first segment of the curve the average velocity is 2.94 mm per half-year, and the average deceleration is 0.26 mm. Over the second segment the average velocity is 1.83 mm per half-year and the average deceleration is 0.48 mm. The third segment of the curve has an average velocity of 0.57 mm per half-year. In contrast to the vectors of velocities for the first two segments of the curve that are highly significant ($p < 0.0005$), that for the third segment is only significant at the $p = 0.094$ level. The multivariate tests and especially the univariate tests indicate that there is no significant horizontal growth of the condyle after 7 years. There is a steady decline in the velocity of growth of this variable over the entire study period. The vectors of decelerations are significant for each of the segments of the curve at better then the 0.05 level.

Superiorly directed growth at the condyle (YCO1, fig. 4.11B) is very strong and constant between 1.5 and 3.5 years, averaging 3.91 mm per half-year. Thus, vertical growth at the condyle is both initially higher (about 1 mm more) than horizontal growth

CONDYLION

A. Horizontal Growth

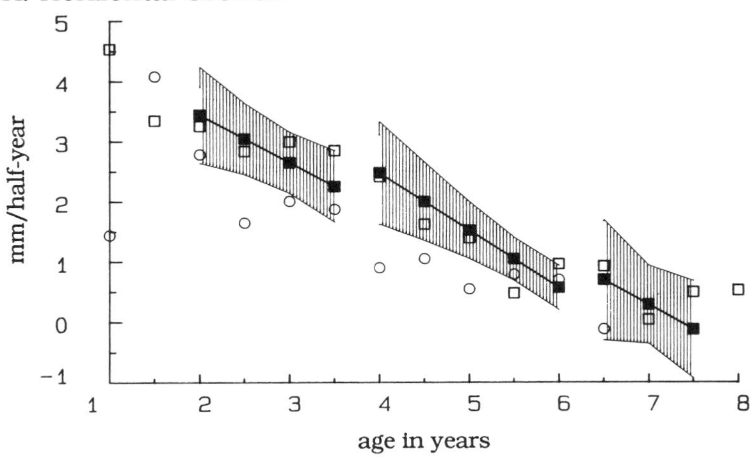

B. Vertical Growth

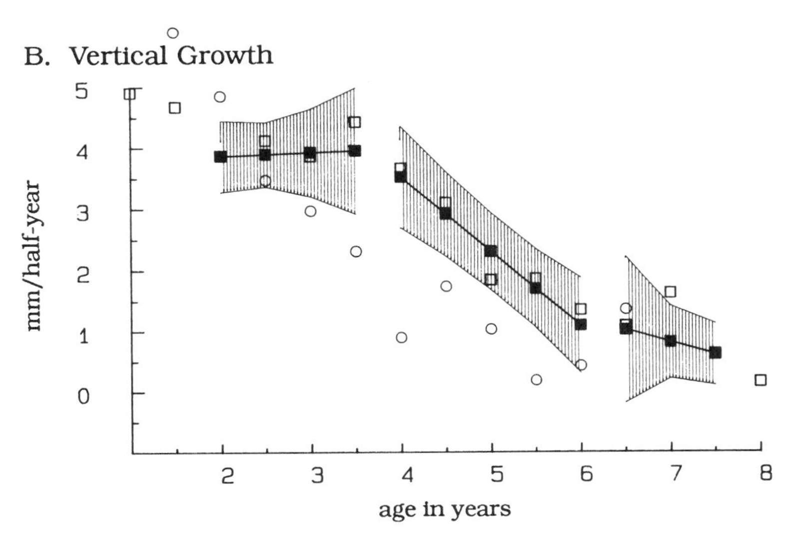

FIGURE 4.11 Rate of growth at *condylion* (X & Y-CO1) in mm per half-year. (A) Horizontal and (B) vertical displacement of this landmark relative to the body of the mandible (MRL).

at this site and becomes proportionately more important over this time period since there is no deceleration. Over the 3- to 6-year period there is highly significant deceleration averaging 0.58 mm per half-year^2 ($p < 0.0005$). It is worthy of note that while the average velocity at the beginning of this segment exceeds the corresponding horizontal condylar growth by about 1 mm, the rate of the former approximates the latter by 6 years because of a greater rate of deceleration. However, in contrast to the horizontal growth over the 6- to 7.5-year period, the vector of velocities over the same period is quite significant ($p = 0.0079$). The univariate means at 8 and 8.5 years suggest that vertical growth ceases soon after 7.5 years. Thus, on average, the vertical component of condylar growth may persist somewhat longer than horizontal growth at condylion.

The direction of growth at condylion (COD, fig. 4.12) is on average, over the whole study period, 47 degrees above the original occlusal plane. From 2.5 to 4.5 years, the mean direc-

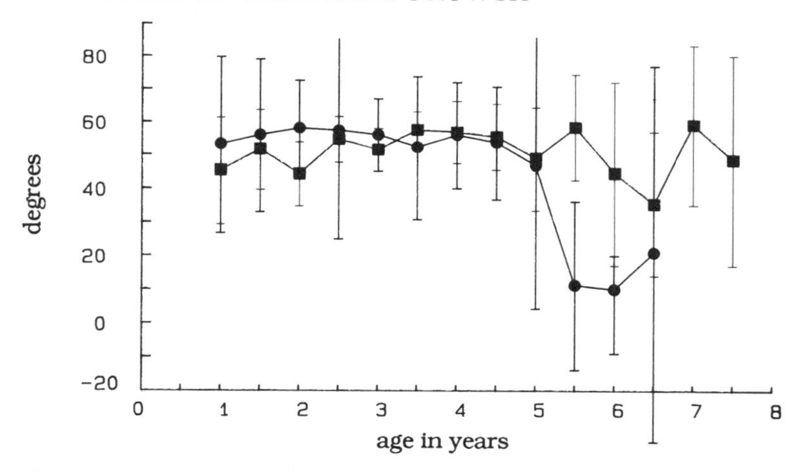

FIGURE 4.12 *Direction of vector of condylar growth* (COD) relative to the body of the mandible (MRL). Mean directions are expressed in degrees. Symbols represent *conventionally* calculated means and 95% confidence intervals.

tion (univariate) is somewhat higher than this overall mean, at 56 degrees. After 6.5 years, direction of growth at this site is extremely variable, as indicated by the very large conventional 95% confidence intervals, but the means tend to fall to approximately 45 degrees in the males.

Remodeling at Infradentale

The anterior component of the repositioning of infradentale, relative to the mandible (XID1), is shown in figure 4.13A. From infancy to about 2 years, the univariate means suggest that supradentale is displacing forward at about 0.8 mm per half-year. The average velocity of growth between 1.5 and 3.5 years decreases to 0.63 mm per half-year. Both the vector of velocities and decelerations are quite significant ($p = 0.0001$ and $p = 0.0053$, respectively) for this period. By 3.5 years there is no significant anterior displacement due to this deceleration of activity. At 4 years there is a clear-cut resumption of horizontal growth here. Over the 4- to 6-year period there is a small but significant and relatively constant quantity of growth, averaging 0.58 mm per half-year. Anterior displacement diminishes over the remainder of ontogeny, as indicated by the significant vector of decelerations over the 6- to 7.5-year period ($p = 0.032$); the growth velocity of 0.16 mm per half-year for this period diminishes to zero by about 7.5 years of age.

Significant vertical displacement of infradentale (YID1, fig. 4.13B) occurs only during the first 3.5 years of life. The average velocity for the 1.5 to 3.5 year interval is 0.50 mm per half-year. Both the vector of velocities and the accelerations for this period are significant at better than the 0.007 level. Examination of the univariate means suggests that this vertical activity might be confined to an episode at around 1 year of age, and another at 3 or 3.5 years. After this period, all velocities and accelerations, when evaluated using both univariate and multivariate criteria, do not differ significantly from zero. The only exception is the

INFRADENTALE

A. Horizontal Growth

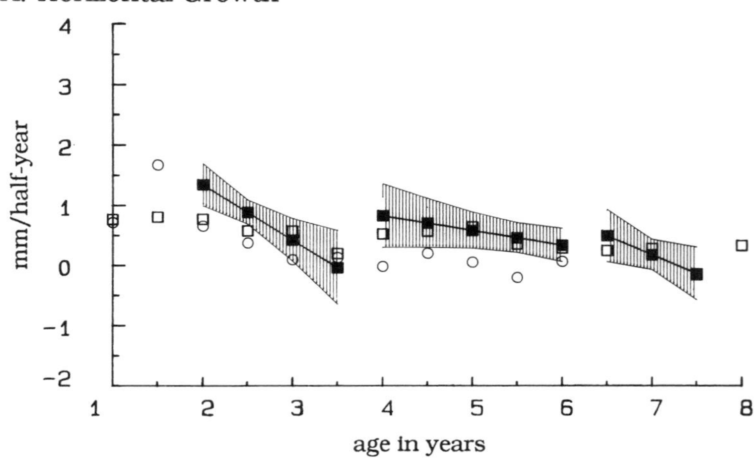

B. Vertical Growth

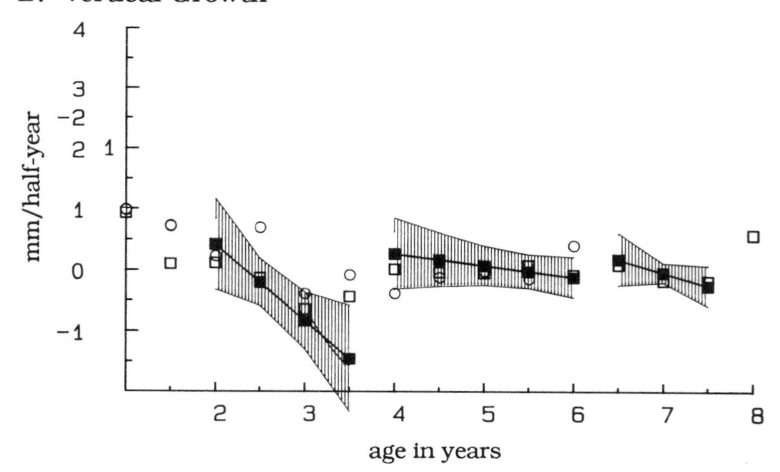

FIGURE 4.13 Rate of growth at *infradentale* (X & Y-ID1) in mm per half-year. (A) Horizontal and (B) vertical displacement of this landmark are relative to the body of the mandible (MRL).

velocity at 7.5 years, which univariately differs from zero (p = 0.016). The confidence bands after 3.5 years clearly demonstrate very modest variability around zero.

Remodeling at Menton

Posterior repositioning of menton (XMN1, fig. 4.14A) occurs until about 6 years of age. During infancy there is little or no horizontal activity here, as indicated by the univariate means. Between 1.5 and 3.5 years there is marginal acceleration of activity at menton, culminating in a growth peak between 3 and 4 years, at which time the average velocity is about 1 mm per half-year. From 4 to 6 years, the rate of horizontal displacement gradually diminishes to zero. Over this period, the average rate of displacement is 0.56 mm; this value and the corresponding vector of velocities are significant at better than the 0.001 level. The average deceleration for this period of 0.16 is significant at the 0.013 level, whereas the vector of decelerations only attains significance at the 0.069 level. At 6 years and beyond, the horizontal position of menton is stable.

Inferior repositioning of menton (YMN1, fig. 4.14B) occurs at an extremely constant and significant average rate of 0.56 per half-year from infancy until 3.5 years. At 4 years there is a peak in the rate at which menton displaces to 0.79 mm per half-year. The segment of the growth curve between 4 and 6 years is distinctly curvilinear, best fit by a quadratic equation. The vectors of velocities, decelerations, and third-order differences are all significant at better than the 0.05 level. Thus, vertical displacement of this landmark is most pronounced at 4 years, decreases to zero by 5 years, and accelerates to a low but significant level of 0.25 mm per half-year at 6 years (p = 0.013 for the univariate mean at this time point). After this time, the position of menton has completely stabilized with regard to both its vertical as well as its horizontal position.

The most notable aspect of the vertical behavior of menton throughout ontogeny is the minimal amount of variability; the

MENTON

A. Horizontal Growth

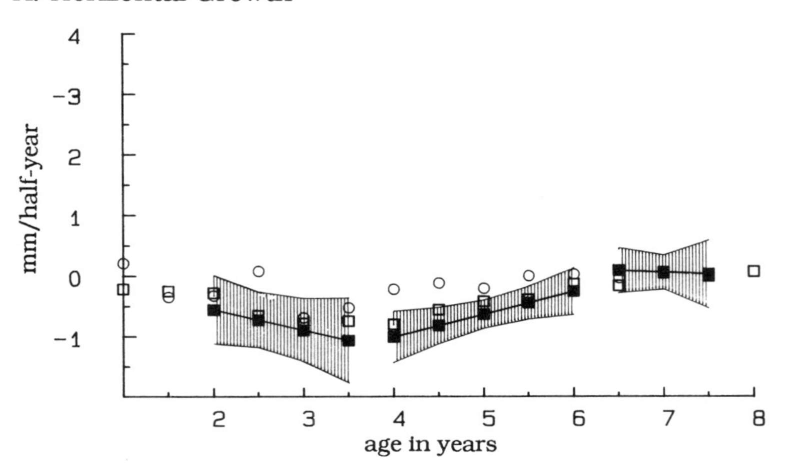

B. Vertical Growth

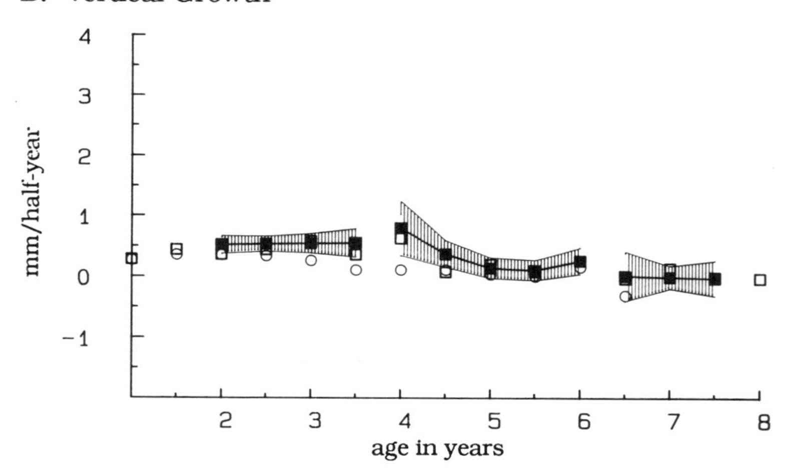

FIGURE 4.14 Rate of growth at *menton* (X & Y-MN1) in mm per half-year. (A) Horizontal and (B) vertical displacement in this landmark are relative to the body of the mandible (MRL).

confidence bands are extremely narrow throughout the study period.

Remodeling at Gonion

Between 1.5 and 3.5 years, the posterior repositioning of gonion (XGO1, fig. 4.15A) occurs at a constant rate averaging 3.77 mm per half-year. There is considerable variability for this segment of the growth curve, and a great deal of variability at 3.5 years. From 3.5 to 6 years the rate of horizontal growth here decelerates significantly; the vector of decelerations is significant at the $p = 0.0001$ level. There is also less variability during this period. The velocity at the nadir of this part of the curve at 6 years of age is 0.69 mm per half-year. Between 6.5 and 7.5 years the evidence of significant growth is marginal. The average velocity for this segment is also 0.69 mm per half-year, which is univariately significant at the 0.018 level, whereas the vector of velocities does not differ significantly from zero. This part of the curve is also characterized by a great deal of variability. The univariate means from 7 to 8.5 years do not differ significantly from zero.

The shape of the velocity curve for the superior displacement of gonion (YGO1, fig. 4.15B) is quite different from that of its horizontal counterpart during the early periods. The univariate and multivariate data from infancy to 3.5 years suggest an acceleration in the rate of superior displacement from zero to a peak of 2.28 mm at 3.5 years; the average acceleration is 0.36 ($p = 0.051$), while the vector of accelerations only attains significance at the $p = 0.12$ level. Between 4 and 6 years there is significant deceleration averaging 0.25 mm per half-year^2; the vector of accelerations is significant at the $p = 0.024$ level. The average velocity for the 6- to 7.5-year period is 0.46 mm per half-year; the vector of velocities for this period is significant ($p = 0.018$), but the accelerations are not. Between 7.5 and 8.5 years the univariate means suggest that gonion is remodeling slightly *downwards*, if at all.

GONION

A. Horizontal Growth

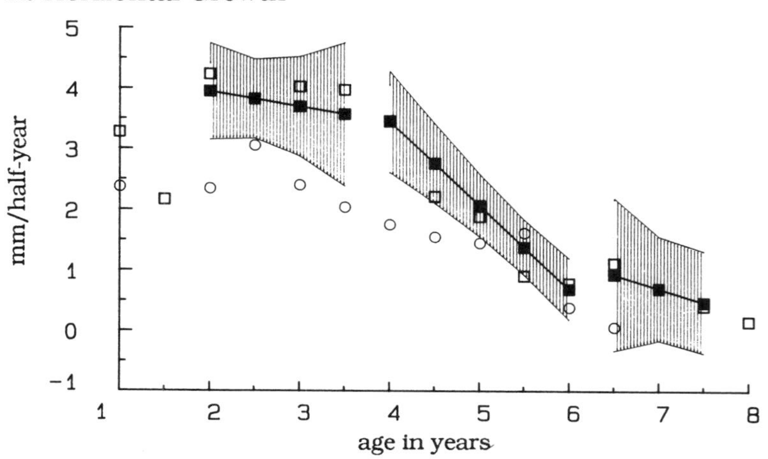

B. Vertical Growth

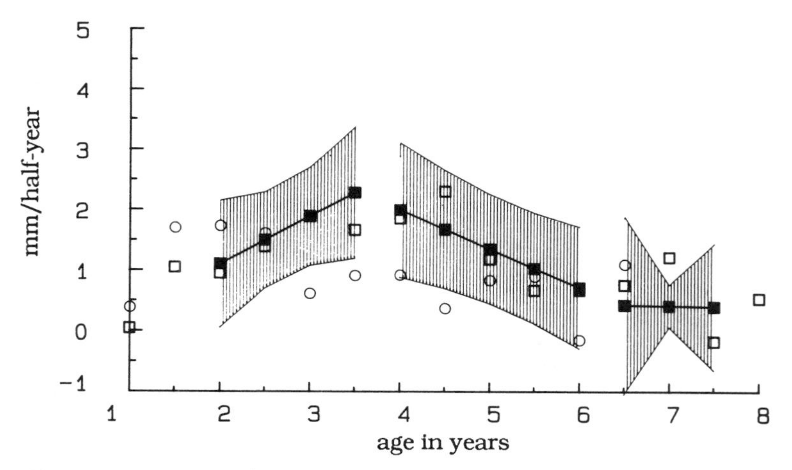

FIGURE 4.15 Rate of growth at *gonion* (X & Y-GO1) in mm per half-year. (A) Horizontal and (B) vertical displacement of this landmark are relative to the body of the mandible (MRL).

Mandibular Alveolar Ridge

Examination of the multivariate curves and the univariate means for superiorly directed growth at the mandibular alveolar ridge in the region of the canine tooth (YMNALVR1, fig. 4.16) reveals that growth here is episodic. During infancy the negligible growth here increases to a peak velocity of 0.96 mm per half-year at 2 years of age. At 3 years growth has again diminished to zero. By 3.5 years, growth has effectively come to a halt. However, by 4 years, activity has resumed and it peaks at about 0.74 mm per half-year. Over the 4- to 6-year period there is significant deceleration (the vector of decelerations is significant at the $p = 0.044$ level). By 5.5 years the growth here is again effectively zero. At 6.5 years there is equivocal evidence of a brief resumption of activity here. The velocity at this time is 0.58 mm per half-year but attains univariate significance only at the 0.079

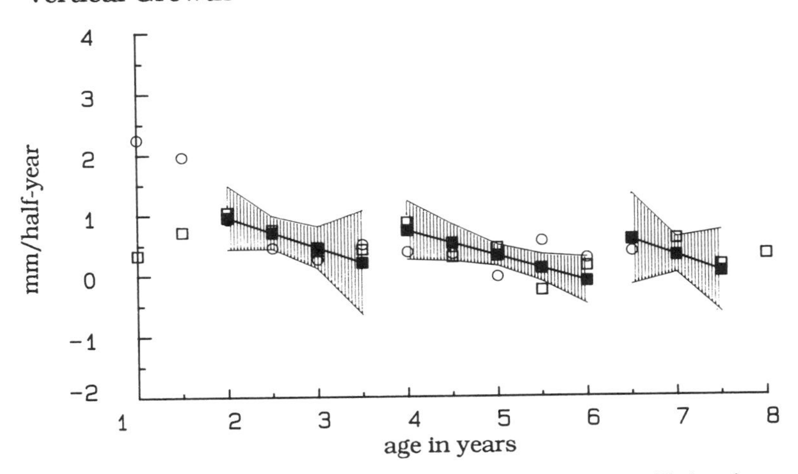

FIGURE 4.16 Rate of vertical (superior) growth at the *mandibular alveolar ridge* (YMNALVR1), that is, the vertical displacement of a point immediately posterior to the canine tooth relative to the body of the mandible (MRL).

level. However, after 6.5 years of age there is only marginal evidence of growth here (for the 6.5- to 7.5-year period the average velocity is 0.31 mm per half-year, p = 0.0042, whereas the vector of velocities attains significance only at p = 0.071). After 7.5 years there is no evidence of growth here.

Mandibular Length

Oblique mandibular length (MNLN, fig. 4.17A) is the most rapidly growing dimension in the facial skeleton of the rhesus monkey. The growth of this dimension is fit well by a linear equation from infancy until about 3.5 years, with an average velocity of 4.65 mm per half-year. The decelerations are not significant. There is minimal variability around the growth curve during this period. Average lengths increase from 51.3 ± 2.69 mm at 1.5 years to 70.6 ± 3.04 mm at 3.5 years.

In infancy the growth velocity of mandibular lengthening (fig. 4.17B) is quite strong. From this time until 3 or 3.5 years of age, the rate first decreases and then increases. At 3.5 or 4 years, there is a growth peak where the velocity attains 4.43 mm per half year. From this point until 6 years there is significant deceleration. Average velocities diminish from 4.43 to 1.36 mm per half-year, causing an average deceleration of 0.86 mm per half-year^2. There is somewhat more variability around this portion of the curve; the 95% simultaneous confidence intervals rise to about 7% of total mandibular length. At 6 years, average mandibular length is 85.6 ± 6.00 mm.

Over the 6- to 7.5-year span, the average growth velocity of mandibular length is reduced to 1.13 mm per half-year. This average increment is significant (p = 0.0025), whereas the vector of velocities is considerably less significant (p = 0.0546). Similarly, the average deceleration of 0.48 mm is significant (p = 0.024), while the vector of decelerations is not (p = 0.0839). These statistics provide an overall picture of marginal increments in this dimension during this period in males. After 7.5 years, the univariate means and the multivariate confidence

MANDIBULAR LENGTH

A. Total Length

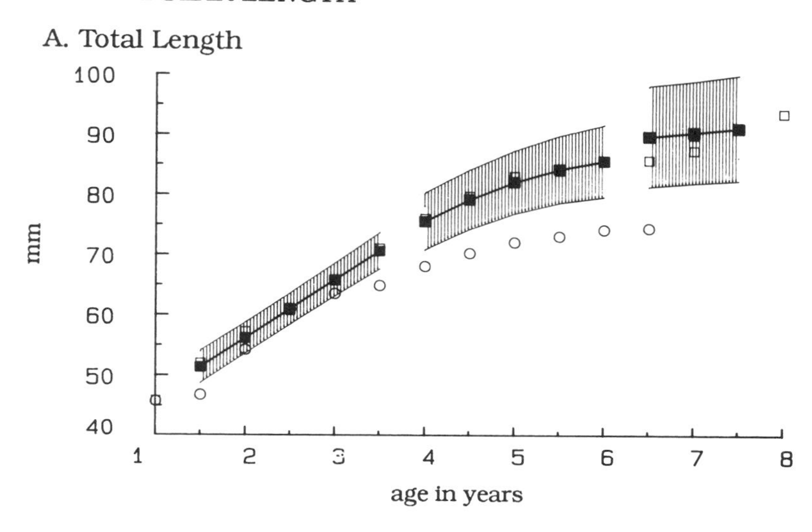

B. Change in Length

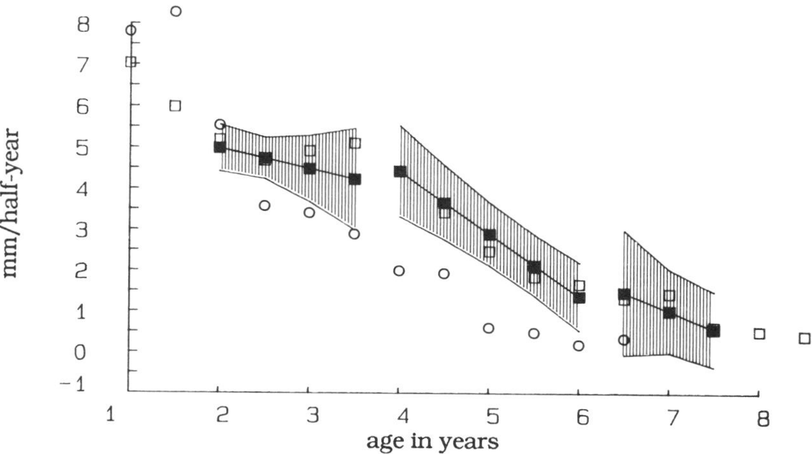

FIGURE 4.17 (A) Growth curve for *mandibular length* (MNLN) as measured obliquely from pogonion to condylion. (B) Growth velocity curve for mandibular length (MNLN1). Increments expressed in mm per half-year.

intervals suggest a stabilization of mandibular length at 93.7 ± 8.77 mm.

Mandibular corpus length (CORPLN) exhibits extremely high correlations with mandibular length. Because of this redundancy in information, data for this variable are not presented here.

Ramus Height

Height of the ascending ramus of the mandible (fig. 4.18), as measured from the most inferior point of the gonial region to the most inferior point along the mandibular notch (RMHT, fig. 4.18A), increases linearly over the first 3.5 years of life. There is very little variability about this segment of the growth curve. At 1.5 years ramus height averages 26.70 ± 0.97 mm. At 3.5 years this dimension averages 36.17 ± 2.30 mm. The average velocity for this period is 2.42 mm per half year. The vector of decelerations is not significant.

The behavior of the variables *ramus height*, RMHT (fig. 4.18A), and *change in ramus height*, RMHT1 (fig. 4.18B) over the next segment of the growth curve is problematic. According to Hills's analysis, the vectors of velocities and accelerations are highly significant ($p < 0.002$), and the third-order differences are also significant ($p = 0.033$). According to Rao's analysis, these data cannot be adequately fit by a polynomial equation of any degree. However, the cubic equation comes closest to meeting the goodness-of-fit criterion so that fitted averages and confidence bands issuing from this equation are presented in figure 4.18 for lack of superior alternatives. At 4 years of age, ramus height averages 39.11 ± 4.37 mm. At 6 years it averages 43.43 ± 5.79 mm. During this period the average velocity is 1.27 mm per half-year and the average deceleration is 0.39 mm per half-year^2.

Between 6 and 6.5 years of age the average growth velocity in ramus height is 0.60 mm per half-year. The vector of velocities over this period is significant ($p = 0.0143$). The vector of decel-

RAMUS HEIGHT

A. Total Height

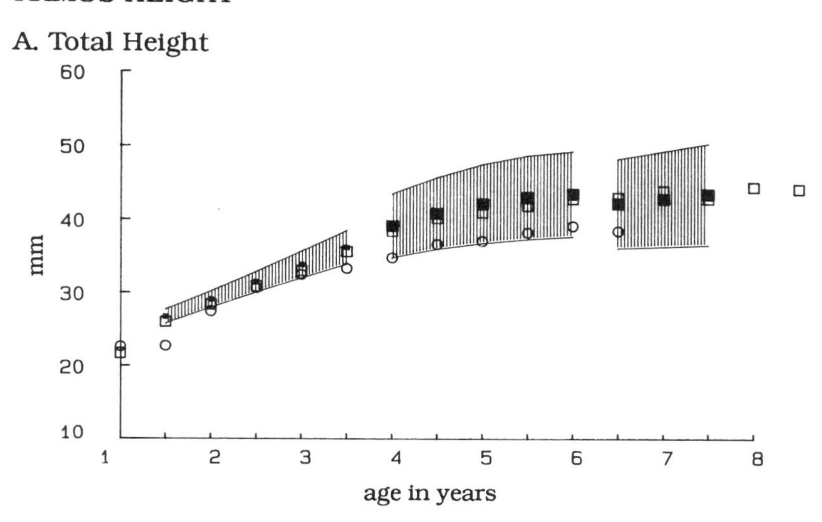

B. Change in Height

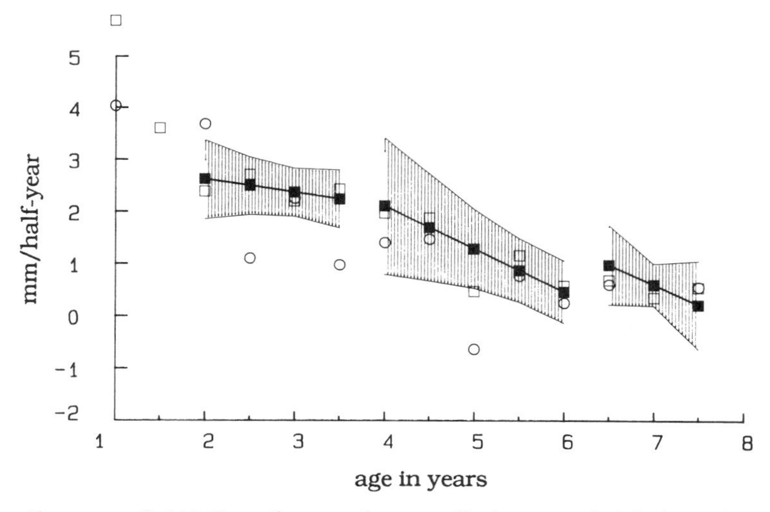

FIGURE 4.18 (A) Growth curve for mandibular *ramus height* (RMHT) as measured from the most inferior point of the ramus to the most inferior point along the mandibular (sigmoid) notch. (B) Corresponding growth velocity curve (RMHT1). The cubic equation of the data for 3.5 to 6 years is portrayed in (A) although it does not meet goodness-of-fit criteria (see text).

erations is only significant at the p = 0.068 level. Over this period there is considerable variability, for instance, at 7.5 years of age ramus height averages 43.69 ± 6.18 mm. The univariate means suggest that ramus height stabilizes after this period.

Summary of Mandibular Growth and Remodeling

The anterior component of mandibular displacement exceeds the inferior component throughout most of the study period. The average rates of anterior displacement for each of the segments of the whole curve are about 50% to 60% greater than the rates of inferior displacement. The shapes of the velocity curves for the two orthogonal components of mandibular displacement are also quite different from one another. While the vertical component of displacement diminishes gradually and almost linearly from about 3.5 mm per half-year to zero by 7 years of age, the horizontal component exhibits a sigmoid curve. The latter curve is characterized by an initially high and constant rate of displacement until 4 years, followed by a marked deceleration until 5.5 years, followed by a plateau in rate over the next 2 years, at which time the rate appears to drop off toward zero.

The inferior component of mandibular displacement plays a proportionately greater role in the posterior part of the mandible than in the anterior region. This is evidenced by the significant counterclockwise rotation that the mandible undergoes relative to the cranial base throughout the first 6 years of life. Thus, while the mandible (or more specifically, any point on the mandible) is *translating* forward and downward, the mandible as a whole (or more specifically, the unremodeled core of the corpus) is simultaneously undergoing *angular translation*, that is, rotating in a counterclockwise direction.

A salient aspect of these results concerning mandibular rotation is that it is always in a *counterclockwise* direction. The velocity curve and 95% confidence band is always negative and well below zero during most of the period of active growth in the craniofacial complex of the male rhesus monkey.

This angular repositioning of the mandibular corpus occurs at the greatest rate when a number of variables reflecting remodeling/bone deposition at various sites in the mandible also exhibit the greatest activity, at about 3.5 or 4 years of age in males.

Posterosuperiorly directed growth at the mandibular condyle accounts for most of the increments in mandibular length, as it has been measured from pogonion to condylion. The resultant vectors of horizontal and vertical displacements of condylion (relative to the mandibular corpus) are roughly equal to total average increments in mandibular length at most observations.

Superiorly directed increments at condylion are greater than posteriorly directed increments at all observations, accounting for the average direction of the growth vector being consistently greater than 45 degrees during most of ontogeny. While horizontally directed growth at this site is clearly decelerating over the first 3 years of life, the vertical component of growth at condylion remains relatively constant and high in magnitude. Thus the direction of the resultant vector of growth here becomes *progressively more vertical* over this period.

Over the 4- to 6-year period, both vertical and horizontal components of condylar growth decelerate at rates that are both comparable and significant. Thus, during this period, there is no evidence for a shift in the *direction* of condylar growth; the vertical contribution continues to predominate. During the 6- to 7.5-year period, the horizontal component of growth here significantly decelerates while the vertical component does not. Thus, the last bit of significant condylar growth that is exhibited over this period comes to be almost entirely vertical. By 8 years of age there is no evidence of any growth at this site.

Counterclockwise mandibular rotation, and the consequent rotation of the MRL, the "horizontal" plane relative to which orthogonal translations above are described, complicates the interpretation of these displacements as this horizontal plane of reference shifts throughout ontogeny. Although the above description of superior growth predominating over horizontal growth at the condyle is qualitatively correct, the actual values reported tend to exaggerate the magnitudes of the vertical components relative to their horizontal counterparts as the animals become older. Since mandibular rotations are greater than those of the maxilla, the artifact is more pronounced in the lower jaw.

Increments at the condyle (specifically, the magnitude of the resultant vector of growth at condylion) and increments in total mandibular length do not exactly sum to each other despite being strongly correlated. These small disparities between a measurement that is determined relative to bone implants and one based on two landmarks can be attributed to (1) growth at a location other than condylion, and (2) geometric considerations. (1) Evidence of modest periosteal apposition of bone along the anterior aspect of the mandibular symphysis is demonstrated quantitatively by the horizontal increments at infradentale, a landmark located superior to pogonion, and graphically by composite plots of mandibles superimposed on mandibular implants. Addition of bone here accounts for 10% to 20% of total increments in mandibular length before 5.5 years of age. (2) The direction of the resultant vector of condylar growth is not always colinear with the line between pogonion and condylion, and will typically differ somewhat in magnitude from increases in the length of this line depending upon the *relative* contribution of each of the two orthogonal components of growth at the condyle itself.

The vertical position of infradentale is quite constant throughout the study period. Only at about 3 or 3.5 years of age does its vertical position change significantly. Menton, on the other hand, undergoes small but significant inferior displacements up until about 4.5 years of age. This inferior displacement again resumes briefly at 6 years of age. Throughout this entire period, menton is also undergoing significant and more pronounced posterior displacement. The repositioning of this landmark represents the thickening of the mandibular symphysis in the anteroposterior dimension, as well as an increase in the height of the symphysis. These results suggest that increases in height of the symphysis of the sort reported in macaques by Sirianni and coworkers (1982) receives a greater contribution from growth in the region of menton than infradentale.

The *shape* of the growth curve for mandibular ramus height resembles that of mandibular length. However, the *magnitudes* of the velocities of growth for this vertical dimension of the mandible tend to be about 40% to 50% of the latter, more oblique measure of mandibular size. It is readily ascertained from the plots of superimposed mandibles that changes in ramus height represent the composite of posterosuperior repositioning of the entire gonial region and of the mandibular notch. Thus, the line representing ramus height is a dimension that is itself, very clearly, repositioning over time.

CHANGES IN THE OCCLUSAL PLANE IN MALES

Changes in the angle of the functional occlusal plane are expressed relative to the maxillary and mandibular bodies. More specifically, the measurements presented below were made with reference to the maxillary and mandibular reference lines that are based on the implants in the basal bone of each of the jaws.

There is only marginal evidence of significant clockwise rotations of the occlusal plane relative to the maxilla during the first 6 years of life (MXRLOCPL1, fig. 4.19A). None of the vectors of velocities or accelerations for each of the three segments of the growth velocity for this variable are significant at the 0.05 level. On the other hand, all of the average values before 7 years of age are above zero, reflecting a *tendency* toward positive, or *clockwise* rotations of the occlusal plane relative to the maxilla. Furthermore, the average velocities for the 1.5- to 3-year and 3- to 6-year segments both attain univariate significance at better than the 0.02 level. In addition, the average de-

OCCLUSAL PLANE ROTATION

A. Change in Angle between Maxilla and Occlusal Plane

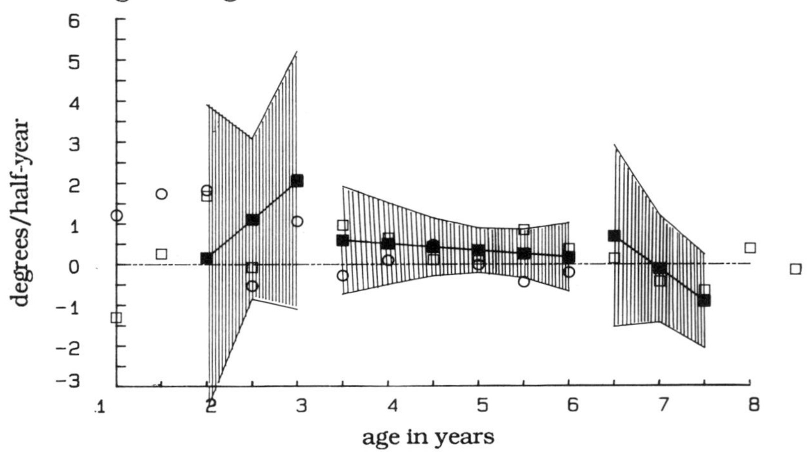

B. Change in Angle between Mandible and Occlusal Plane

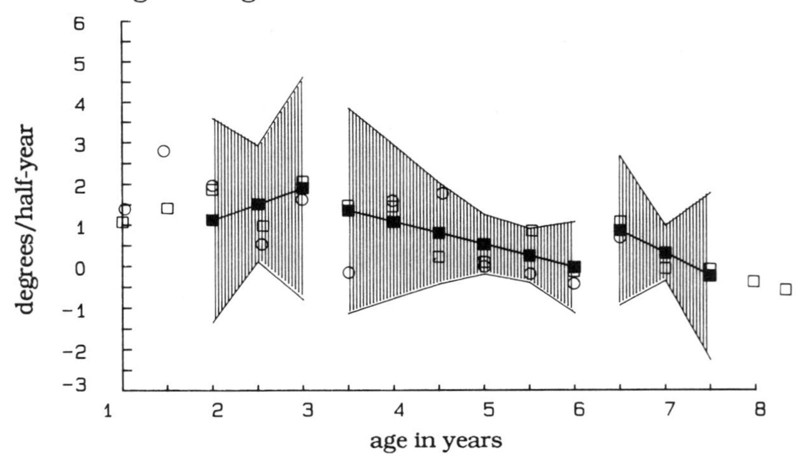

FIGURE 4.19 (A) *Maxillary occlusal plane rotation* (MXRLOCPL1), rate of rotation of the functional occlusal plane relative to the body of the maxilla (MXRL); and (B) *mandibular occlusal plane rotation* (MRLOCPL1), relative to the body of the mandible (MRL).

celeration for the 3- to 6-year span is significant at the 0.012 level. Examination of individual univariate velocities suggests that there is a peak in the rate at which the occlusal plane is reorienting relative to the maxilla at about 3 years ($p = 0.0083$) and at 5.5 years ($p = 0.0084$). During the first 3 years of life there is enormous variability in this measurement.

The occlusal plane exhibits significant clockwise rotation relative to the mandible over the first 3 years of life (MRLOCPL1, fig. 4.19B). The vector of velocities over the 1.5- to 3-year period is significant at the 0.02 level. The average velocity over this period, during which there is marked variability, is 1.67 degrees per half-year. The accelerations over this period are not significant. Over the 3- to 6-year period, the vectors of velocities and decelerations only attain significance at the 0.085 level. The only velocities to attain univariate significance are at 3.5 and 5.5 years of age ($p = 0.0125$ and 0.026, respectively). The average velocity and deceleration over the 3- to 6-year period attains univariate significance at the 0.0011 and 0.012 levels, respectively. Similar to the behavior of the occlusal plane relative to the maxilla, all of the average values for this variable until 7 years of age are above zero (except at 6 years), suggesting a consistent trend of rotation in the clockwise direction. The shapes of the curves are also very similar, though the average magnitudes of the changes relative to the mandible are slightly elevated above those relative to the maxilla until about 5 years of age.

Summary of Angular Changes in the Jaws and Occlusal Plane

Both jaws exhibit very significant *counterclockwise* rotations relative to the cranial base over the first 6 years of life. The rate of rotation is somewhat greater in the mandible than in the maxilla. Both jaws demonstrate peaks in the rate of rotation at about 3.5 years and possibly again at 5.5 years. The measures of rotation of both jaws are characterized by considerable variability.

The occlusal plane clearly tends to rotate in the *clockwise* direction relative to both jaws. The velocity curves and confidence bands of the measures reflecting the reorientation of the occlusal plane relative to the jaws correspond closely in shape to the angular changes that the jaws exhibit relative to the cranial base, though the directions of these angular changes are opposite. Furthermore, there is evidence of peaks of rotational activity in both the jaws *and* occlusal plane at 3.5 and 5.5 years of age. These changes in the orientation of the occlusal plane appear to offset or compensate for the tendency of the bodies of the jaws to rotate counterclockwise relative to the cranial base.

COORDINATION OF MAXILLOMANDIBULAR GROWTH

Correlations among the change variables provide an indication of how activity at the various growth sites is coordinated (fig. 4.20). First, correlations between maxillary and mandibular variables were examined. Posterior growth at the two maxillary tuberosity sites (maxillopalatine junction [MXPJ] and pterygomaxillary fissure point [PTMXFI]) and increments in total maxillary length are strongly correlated with posterior growth at all three condylar sites (condylion: CO; superior and posterior condylion: SCO, PCO), increments in mandibular length ($r > 0.45$, $p < 0.0001$), and to a lesser extent, vertical growth at the three condylar sites and increments in ramus height ($r > 0.37$, $p < 0.0003$) during the first 3.5 years in males. Increments in maxillary length exhibit highly significant correlations with these same mandibular change variables over the same period in males ($r > 0.52$, $p < 0.0001$). These are the strongest class of correlations between maxillary and mandibular variables. Numerous other combinations of variables have correlations in the 0.15 to 0.30 range that are also highly significant. However, attention has been limited to those relationships which appear to be the most profound.

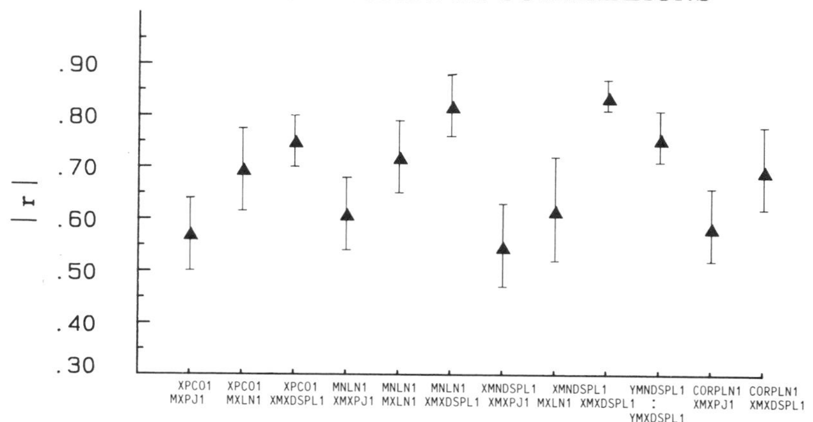

FIGURE 4.20 Pearson product moment correlations between various maxillary and mandibular change variables, that is, rates of growth and displacement. The pairs of variables are indicated along the abscissa. Ninety-five percent confidence intervals for the coefficients are also shown. Both sexes are combined, and all observations are treated together. The set of correlations is limited to those where $p(r = 0) < 0.05$.

During the 4- to 6-year age period a very similar pattern of correlations are observable in the males. This lack of change in the pattern of correlations is supported by the finding that the confidence intervals for the correlation coefficients for this second part of the life span almost completely overlap those of the earlier phase. Over the 6.5- to 7.5-year period, a number of the correlations cease to be significant. For example, the correlations between horizontal growth at the maxillopalatine junction point (XMXPJ) and vertical and horizontal growth at condylion, horizontal growth at superior condylion, and increments in mandibular ramus height. After 7.5 years all of the correlations considered above cease to be significant except those between vertical changes at superior condylion and changes in ramus height.

Remodeling of the gonial region is significantly correlated

with growth at the condyle during the first 3.5 years of life in males. The strongest correlations are between the horizontal displacements of gonion and both vertical and horizontal displacements at the three condylar sites ($0.61 < r < 0.77$, $p < 0.0001$). Equally significant correlations, but of smaller magnitude, are present between the vertical displacement of gonion and the vertical displacements of the three condylar landmarks ($0.42 < r < 0.46$).

The displacements of the jaws are highly correlated with one another. The coefficient for the horizontal displacement of the mandible with that of the maxilla was the highest (and was the most significant of all of the correlations examined). The correlation between the vertical displacement of the mandible with that of the maxilla was also quite strong. Thus, coordination in the displacement of the jaws is more pronounced than that occurring between at any two sites of growth, for example, those described above. The coefficients portrayed in figure 4.20 are somewhat higher than those discussed above, as they are not broken down by time intervals or sex; all observations have been treated together.

SEXUAL DIMORPHISM

Maxilla

Absolute maxillary dimensions (figs. 4.7a–4.8a) are indistinguishable between males and females until 3.5 years of age; the female mean lengths and heights (calculated conventionally) are almost identical to the corresponding male (multivariate) averages over this period. Maxillary height continues to be very similar between the sexes throughout the rest of the study period. On the other hand, the disparity between average male and female maxillary lengths becomes greater throughout the remainder of the study period. The average male maxillary length exceeds that of the female by about 11% at 3.5 years, and by about 16% by 7.5 years of age.

The rate of increase in maxillary length (MXLN1) in females is similar to that of the males at 2 years, but afterwards it decreases so that it is at the lower border of the male confidence band until about 5.5 or 6 years. At this time the rates of maxillary lengthening appear very similar.

Two sample *t*-tests, using Bonferroni corrections for multiple tests, indicated no significant sex differences in overall maxillary dimensions in all animals under 3.5 years old. This test on animals 3.5 to 6 years old, for which both sexes are well represented, indicates significant differences in maxillary length ($p < 0.0001$) but not in maxillary height.

The rate of anterior maxillary displacement (XMXDSPL1) is consistently greater in the males than in the females from 2.5 until about 6 years of age, at which time the rate is effectively zero in both sexes. During infancy, the female rate of anterior displacement may be greater. The disparity in rates between the sexes is greatest between 2.5 and 5 years. The mean rates of inferior displacement of the maxilla (YMXDSPL1) for the female sample fall well within the narrow confidence band for the male growth curve. The shapes of the velocity curves for vertical maxillary displacement are quite similar between the sexes; the main difference is one of magnitude. The velocity curves for horizontal maxillary displacement tend to differ between the sexes in both shape and magnitude.

The pattern of maxillary rotation suggested by the female data appears somewhat different from that of the males during the early years. In the infant males, the negligible rotation increases to a substantial average rate of counterclockwise rotation by 3 or 3.5 years, followed by a first precipitous and then gradual decline in the rate of rotation, ending at about 6.5 years of age. In contrast, maximum counterclockwise rotation is observed in the females during infancy; this rate apparently decreases steadily to zero by 3.5 years of age.

The female pattern of growth at the maxillopalatine junction point (MXPJ, fig. 4.3) is indistinguishable from that of the males, except perhaps for a slightly lower horizontal rate of displacement between 3.5 to 5 years; the female means are well within the male confidence band throughout the study period.

The average horizontal growth at the pterygomaxillary fissure inferior point (XPTMXFI1, fig. 4.4) is indistinguishable between the sexes, except perhaps at 2.5 years, when the female rate falls below the male confidence band. An average growth curve and a 95% confidence band was calculated for the smaller female sample over the 2- to 6-year period. The male confidence band is almost completely subsumed by the female band during this period.

The average vertical displacements at the pterygomaxillary fissure point in the females exceed the male values during infancy, but are lower than the male average growth curve afterwards, until 3 years of age. However, only between 2.5 and 3 years of age do the female values fall outside (below) the broad male confidence bands. Univariate evidence of significant vertical growth at this site in the females is confined to activity at 1.5 to 2 years, at 3.5 years, and again at 4.5 years of age; the female means approximate zero at the other observations.

The very marginal changes at supradentale in the females are confined to significant anteroinferior displacement at 2 years of age and purely anterior displacement at 3 years of age. This contrasts the male pattern of *significant*, but low level anterioinferior (predominantly anterior) displacement of this landmark up until 6 years of age. A similarly dimorphic pattern is seen in which the rate of vertical changes at the alveolar ridge (YMXALVR1) is slightly higher in the males.

Mandible

The pattern of sexual dimorphism for total mandibular length (MNLN, fig. 4.17a) is the same as that seen for maxillary length (MXLN, fig. 4.7a); absolute lengths are almost identical until 3.5 years at which time a widening sexual disparity begins. Mandibular length appears to plateau in the females at about 6 years and perhaps not until 7.5 or 8 years of age in the males, paralleling the maxillary pattern.

Total ramus height is also virtually identical between the sexes until 3.5 years, at which time a small disparity of about 7% becomes apparent. At 4 years this dimension in males exceeds that in the females by about 11% and remains constant at this level throughout the rest of the study period.

Two sample *t*-tests using Bonferroni corrections indicated no significant differences in total mandibular length, corpus length and ramus height in animals under 3.5 years old. During the 3.5- to 6-year age period all three of these mandibular dimensions are significantly greater in the males ($p < 0.0001$).

Horizontal displacement of the mandible (XMNDSPL1) is a highly dimorphic variable. The rate of anterior displacement is strong and reasonably constant in magnitude over the first 4 years of life in the males. In contrast, the female means, which are considerably below the male values between 2 and 5 years of age, undergo a steadily declining rate of displacement throughout the study period. Thus, while the male growth velocity curve is clearly sigmoid in shape, the female values suggest a more linear trend. There is evidence of a resurgence of anterior displacement in the males at 6 years of age, which is paralleled by a peak among the females about one half-year earlier. While there is no evidence of horizontal mandibular displacements in the females after 6 years of age, the males continue to exhibit significant anterior displacement until at least 7.5 years of age.

The mean vertical displacements of the mandible (YMNDSPL1) in the females fall below or are at the lower border of the male confidence band between 3 and 5.5 years. Before and after this period, the rates are quite similar between the sexes. Vertical displacement of the mandible ceases in the females by 4 years of age, approximately a half-year earlier than in the males.

The pattern of dimorphism in mandibular rotation relative to the cranial base is very similar to that seen in maxillary rotation.

The highest levels of counterclockwise rotation of the mandible in the females are found in infancy. From this time on, the rate of mandibular rotation steadily declines in the females so as to be effectively zero by 5 years of age. This contrasts the male pattern, in which there is a pronounced peak in the rate of rotation at 3.5 or 4 years of age, followed by a decline, with a nadir at 6.5 years of age, followed by additional significant rotation at about 7.5 years.

The rate of horizontal mandibular condylar growth may be greater in females than males at 1.5 years of age. From this time until 5.5 years of age, the rate of growth at this site in the males is considerably greater than in the females; after this phase the rate of growth at this site is indistinguishable between the sexes.

The rate of vertical condylar growth appears to be greater in the females than in the males at the 1.5- to 2-year observations. After this time it steadily decreases in the females so as to be consistently much below the male confidence band until 6 years of age. Whereas the activity at this site in the females is effectively zero by 5.5 years, it does not decrease to these levels in the males until about 8 years.

A similar pattern of dimorphism to that observed at the condyle is found at the gonial region. The greatest disparity in growth rates between the sexes appears at both sites at 4 years of age.

5

DISCUSSION

MORPHOLOGICAL ISSUES

A number of studies on normal craniofacial growth have been conducted on the rhesus monkey over the years using a variety of techniques and sampling designs, leading to a somewhat disjointed picture. Most of these studies have been limited to short periods of the lifespan, and have had to combine sexes due to small sample size. Furthermore, the sex of animals used in these studies has frequently been unreported or ignored.

The following discussion attempts to integrate information from these earlier studies with the present findings and address the subtleties of age and sex related changes. The various components of the craniofacial complex are first discussed individually and then considered together as they pertain to the topics of (1) the adolescent growth spurt, (2) sexual dimorphism, (3) jaw rotations and compensatory mechanisms, (4) condylar cartilage growth, (5) masseter growth, and (6) differences between the rhesus monkey and humans. The focus of this discussion is on normal mixed-longitudinal cephalometric studies. The consideration of histological studies is limited to those studies which bear on changes in craniofacial morphology at the gross level. The various statistical techniques employed in this study are discussed after these morphological issues.

Maxillary Complex

The present findings confirm and provide quantitative evidence for Enlow's (1966) description of maxillary growth that was based on a cross-sectional histological study of young rhesus monkeys. Deposition of bone at the posteriorly facing maxillary tuberosity is strongly related to the anterior repositioning as well as the anteroposterior enlargement of the maxillary complex that occurs until adulthood. Enlow's picture is further refined by the observation that the addition of bone at the pterygomaxillary fissure inferior point is not strictly posterior; the vector of growth here averages 22 degrees above horizontal. Thus, a superior component to the vector of growth at this tuberosity site (and perhaps elsewhere) plays a significant role in the inferior component of maxillary displacement. These findings contrast Baume's (1951a) assertion that growth at the maxillary tuberosity and premaxillary suture contribute *equally* to increases in maxillary length. Premaxillary sutural growth clearly plays a secondary role in overall lengthening.

Periosteal deposition of bone along the facial aspect (*lamina externa*) that is more or less continuous throughout subadulthood also contributes significantly to enlargement of the maxilla in the anteroposterior dimension. The sagittal depth of the maxillary arch is increased by the gradual deposition of new bone along the entire free margin. Vertical growth at the alveolar ridge contributes about 7 mm to overall maxillary height. This tends indirectly to support Baume's (1951a) observation that growth at the alveolar crests and cribiform plates of the alveolar process makes a relatively small contribution to the vertical development of the maxilla relative to sutural and periosteal bone formation at the orbits and palate, respectively. Additional measurements based on maxillary body bone implant superimpositions must be taken to delineate the sources of maxillary enlargement more fully, particularly in the vertical dimension.

The low magnitude but relatively *constant* depository activity found at supradentale and the alveolar ridge suggests that periosteal growth in these regions may be under the influence of different control mechanisms than that occurring at the maxillary tuberosity.

The results of this study also confirm and refine the quantitative findings of McNamara (1972) and McNamara, Riolo, and Enlow (1976) on maxillary growth of the rhesus monkey. Horizontal displacement and growth of the maxilla predominates over the vertical at all ages studied, resulting in a vector of growth that is consistently less than 45 degrees relative to the cranium. This contrasts with the findings of Gans and Sarnat (1951) and Enlow (1968) with regard to the dominance of vertical over horizontal growth in juvenile and young adolescent monkeys.

As in the human (Björk 1966), the direction of maxillary displacement is quite variable. In contrast, however, the vertical component of displacement is proportionately greater in the human, yielding a resultant vector with a mean direction of 51 degrees (Björk 1966), compared to 10 degrees in the rhesus monkey.

Mandible

The descriptions of Enlow (1966), Elgoyen and coworkers (1972), and McNamara and Graber (1975) of normal mandibular growth in the rhesus monkey are largely supported by the present findings. In agreement with these studies, appositional growth on the anterior (labial) surface of the mandibular body *does not* play a major role in the lengthening of the mandible, as claimed by Baume (1951b). Observations of McNamara and Graber (1975) concerning remodeling along the labial and inferior aspects of the mandibular corpus and anterior and posterior borders of the ramus are confirmed by the composite plots of the present data.

As reported by McNamara and Graber (1975), the ramus be-

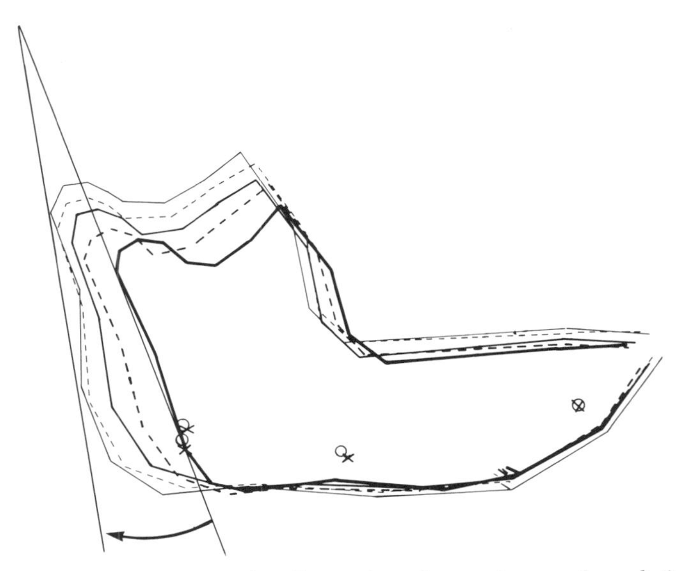

FIGURE 5.1 Composite plot illustrating change in angular relationship between mandibular ramus and corpus. The serial plots are superimposed on mandibular bone implants.

comes more vertical relative to the corpus with maturation. That is, the angular relationship between the posterior ramal border and the body of the mandible becomes slightly more acute with age (fig. 5.1). It can be readily seen that this apparent posterior "relocation" of the ramus is due to differential remodeling along the ramal borders, pronounced posterosuperior remodeling of the gonial region, and a typically greater vertical component to the direction of condylar growth toward adolescence (see below).

McNamara and Graber (1975) illustrated that growth rates at the condyle, ramal borders, and anterior and inferior borders of the corpus decrease in each successive age group. Though rates of growth are relatively high during the infant and juvenile periods (less than 2.5 years old), strong but declining rates of growth after this age continue to make major contributions to adult size. For instance, about 60% of adult male mandibular length is added *after* 2.5 years of age.

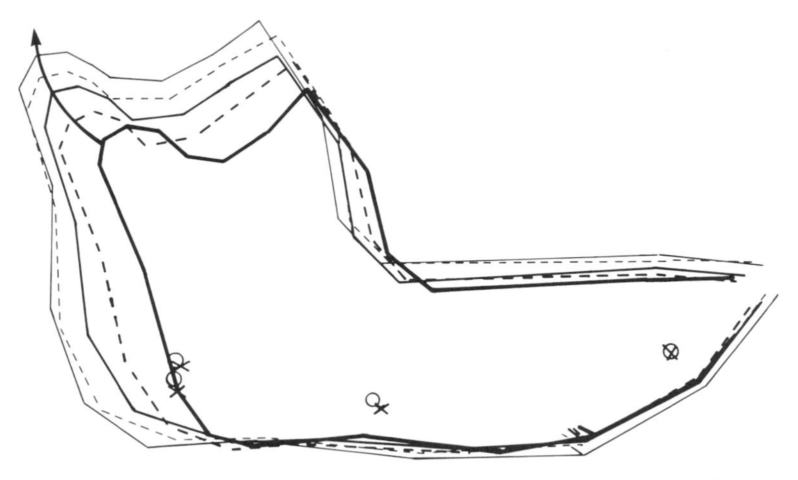

FIGURE 5.2 Composite plot illustrating the anterior curvature of the condylar growth trajectory. Superimposition is on implants in the mandibular body.

McNamara and Graber's (1975) report of a greater horizontal contribution to the posterosuperior vector of condylar growth in the transition from infant to juvenile is supported by the present univariate data. However, findings from later periods indicate that the vertical component consistently predominates, though not to a great extent, over the horizontal component of growth at the condyle. The vertical component plays the greater role prior to adolescence, between 2 and 4 years of age in males. An anteriorly directed *curvature* to the trajectory of condylar growth (fig. 5.2)—of the sort described by Björk (1963) in humans—is seen in many of the monkeys during the periods of most active growth; this is a reflection of this change in the direction of condylar growth.

Adolescent Growth Spurt

The majority of craniofacial dimensions and sites of growth examined in the males in this study exhibit growth velocity peaks at 3.5 to 4 years. However, most of these peaks do not demonstrate accelerations that are statistically significant. That is, the rate of growth (velocity) does not increase significantly (accelerate) during this period. In fact, very few accelerations calculated for any variables are positive, and, of these, fewer are significant. Thus, these data do not provide direct evidence for an adolescent growth spurt. Adolescent growth spurts have not been definitively identified in other craniofacial growth studies in the rhesus monkey (Sirianni and Swindler 1979), nor in the chimpanzee (Gaven 1953).

A possible explanation for these finding is that, due to individual variability in the onset of short-lived accelerations in growth, they are not reflected in the estimated population parameters. Moreover, the technique for filling in missing data may have dampened rate-change trends (both accelerations and decelerations). Larger, more complete samples may be required to address this problem adequately. Better methods for estimating developmental age may also contribute to our understanding of adolescence in nonhuman primates. A scale of maturational status that is derived independently of chronological age, perhaps a multivariate index that reflects proportional or shape changes, may permit the grouping of animals so that a common phase of accelerated growth would emerge.

Aligning growth velocity curves for the individual animals on peak velocities, as is commonly done with human longitudinal data (Israelsohn 1960; Tanner 1962) would almost undoubtedly have provided "evidence" of an adolescent growth spurt. This procedure was not used because this study has sought to describe craniofacial growth as a function of chronological age.

These data from rhesus monkeys, when compared to findings on cranial base growth (Sirianni and Van Ness 1978) and mandibular growth (Sirianni et al. 1982) in the pigtailed macaque (*Macaca nemestrina*), suggest that growth rates peak somewhat earlier in the rhesus monkey than in the larger but very closely related species. The present results indicate a fairly systematic growth peak in *male* rhesus monkeys at about 3.5 to 4 years versus 4 to 5 years of age in the pigtailed macaque males.

Using a less stringent criterion, Sirianni and coworkers (1982) demonstrated an adolescent growth spurt in the mandible of the pigtailed macaque. Deviations of each individual animal's actual growth curve from its fitted polynomial regression curve indicated a consistent pattern of negative deviations (residuals) during the juvenile period, and positive deviations during adolescence. Such growth spurts in males were also reported to be greater in magnitude and duration in the males than in the females. Earlier, this approach had been successfully used to demonstrate adolescent growth spurts in the long bones of chimps (Watts and Gaven 1982) and of rhesus monkeys (Watts 1980).

Though the present results do not support the presence of a systematic acceleration in growth during adolescence, there is strong statistical evidence for a significant deceleration in rate for the majority of measurements that exhibited strong growth during the preadolescent phases in males. Such decelerations after 3.5 years are exhibited in the anteroinferior displacements of both jaws, posterosuperior growth at the maxillary tuberosity and at the mandibular condyle, overall length of both jaws, overall ramus height, the rate of mandibular rotation, inferior remodeling of menton, posterosuperior remodeling of gonion, and vertical increases in mandibular alveolar ridge height.

Though based on graphic impressions rather than rigorous statistical tests, the limited female data suggest a more gradual decline in growth rates in most measurements from infancy to adulthood.

Sexual Dimorphism

At 6 years of age, the last time point for which we have an adequate number of female observations, the males average 13% to 16% larger than females in sagittal jaw dimensions and 7% to 10% larger in vertical dimensions of the jaws. Though these data have limitations, they suggest that sex differences in average growth rates at a number of sites account for the pronounced sexual dimorphism in overall dimensions of both jaws.

The continuous distribution of observations in our data set

provides indications of sexual dimorphism that could not be detected in earlier studies on craniofacial growth rates in rhesus monkeys because of the nature of the available sample (McNamara and Graber 1975; McNamara et al. 1976). In these earlier studies, the female sample was lacking during that part of the life span in which sexual dimorphism is apparently greatest, between the juvenile and young-adult periods.

In most growth rates, the sexes are either identical or the females are growing somewhat faster until 2 years of age. From approximately 2.5 to 4 years of age, the female rates of growth steadily decline, while the male rates are sustained at higher levels. Aspects of maxillary growth and repositioning that behave in this dimorphic fashion are anterior displacement, overall lengthening, and rotation of the maxilla, as well as posterior growth at the inferior maxillary tuberosity site (maxillopalatine junction point), superior growth at the more superior site (pterygomaxillary fissure, inferior point), and inferior growth of the maxillary alveolar ridge. Some of these apparent dimorphisms in growth rates persist until 5 years of age.

This same pattern of dimorphism observed in the maxilla is also very apparent in many of the mandibular measures, but persists until 5 or 6 years of age in more of these variables. This trend is very pronounced in anteroinferior displacement and overall lengthening of the mandible, mandibular rotation, posterosuperior growth at the condyle and at gonion.

The *direction* of growth at the maxillary tuberosity is indistinguishable between the sexes at most observations. In addition to a great overlap of ranges, the means are quite similar. One possible exception to this uniformity occurs at 3 years of age, at which time the average direction is considerably more vertical in the males.

Vertical growth at supradentale in the maxilla, and at infradentale and menton in the mandible, appears to exhibit dimorphism similar to that found in the more rapidly growing sites above. However, its behavior is not as clear-cut during the mixed dentition stage.

Unlike the sexual difference in the *magnitude* of the rate, the average *direction* of condylar growth is remarkably similar between the sexes until 5.5 years, when the marginal condylar growth in females is almost entirely horizontal. The mean direction is almost identical at all but two of the first ten half-year observations. After 5 years the direction of growth in the males continues to average somewhat greater than 45 degrees, except at 6 and 6.5 years, when it averages about 40 degrees.

The pattern of dimorphism observed at the condyle was also present in the gonial region. In both regions, the greatest disparity in rates of growth between the sexes is at about 4 years of age. Carlson (1983) demonstrated the concerted growth and repositioning of the masseter muscle and the gonial region in young rhesus monkeys by means of radiopaque muscle as well as bone implants (see detailed discussion below). Together these findings suggest that growth of this musculoskeletal complex may be under significant hormonal influence during adolescence.

As in the human, these data suggest that there is much overlap between males and females for many longitudinal measures of craniofacial growth (Riolo et al. 1974). Because of this significant overlap, it may be difficult to demonstrate statistically significant differences, even with somewhat larger samples. Nevertheless, all of these trends toward dimorphism in growth rates contribute to significantly greater dimensions of the jaws in male rhesus monkeys, giving the characteristic appearance of larger muzzles.

Jaw Rotations and Compensatory Mechanisms

The general finding of counterclockwise rotation of the jaws during puberty in humans (Björk and Skieller 1972) and in monkeys (McNamara et al. 1976) is paralleled by the present findings in the rhesus monkey. These results indicate that the pattern of counterclockwise rotation occurs throughout the growing period from infancy until about 6 years in males and 5 years in females. The maximal amount of rotational activity occurs in the males at 3.5 to 4 years, though it is not as strongly correlated with condylar growth in the monkey as in the human. As in the human (Björk and Skieller 1972), mandibular rotation in the rhesus monkey is greater in magnitude and more variable than maxillary rotation.

The counterclockwise rotation of both jaws appears to be a very real and pervasive phenomenon in the rhesus monkey. No individuals in this entire longitudinal sample of thirty-five monkeys exhibited average clockwise changes in the angulation of either jaw relative to the cranial base. This finding is in agreement with the rhesus monkey findings of McNamara and Graber (1975) and McNamara and coworkers (1976). This suggests that clockwise rotations of the jaws in the rhesus monkey over any extended period of time would be aberrant.

As with Björk and Skieller's (1972) findings in the human, and those of McNamara, Riolo, and Enlow (1976) in the rhesus monkey, counterclockwise rotations of the basal portions of the jaws tend to be obscured by dentoalveolar changes and remodeling along the periphery of these bones. In addition to illustrating a basic biological phenomenon, this observation also demonstrates the analytical risks involved in attempting to describe the dynamics of facial growth (in particular the motions of the various components relative to each other) when measurements are based solely on landmarks that are themselves repositioning.

Changes in the orientation of the occlusal plane relative to the cranial base are minimized by eruptive compensations that occur while the jaws are rotating. When the occlusal plane is evaluated relative to the maxillary and mandibular bodies, it is found to be rotating in the opposite, or clockwise, direction, offsetting the net effect of the jaw rotations. Thus, the overall shapes of the jaws and their dispositions relative to each other and the cranial base are more or less maintained throughout ontogeny over long periods of time, despite the fact that they are largely comprised of new bone and house a new set of teeth by adulthood.

These findings tend to indicate in the rhesus monkey the presence of the *dentoalveolar compensatory mechanism* postulated to exist in the human by Solow (1966). The postulation that the rotations of the occlusal plane (assessed relative to both the maxillary and the mandibular bodies) are *compensatory* rather than *primary* is supported by the observation that these measurements exhibit considerable variability during periods of vigorous growth. Compensatory remodeling of the basal portion of the mandible to maintain a more or less constant mandibular shape (cf. Enlow et al. 1971) is to a great extent confined to the gonial region and ramus.

In contrast to the pattern of mandibular growth observed in the human (Björk 1963), there is minimal remodeling of the corpus in the rhesus monkey after approximately 6 months of age (fig. 5.3). The present evidence suggests that mandibular growth, at least as viewed in the lateral projection, appears to be a structurally very conservative process; apart from the anterior border of the ramus, there are no major areas in which the amount of bone loss approximates the amount of gain observed at the major sites of deposition. We must recognize, however, that this interpretation is based on the limited two-dimensional

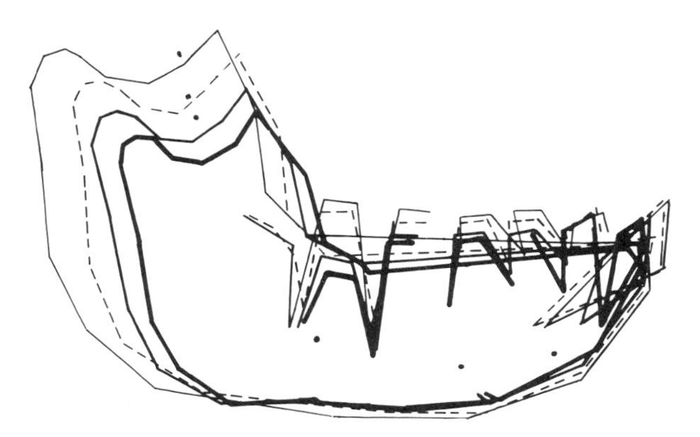

FIGURE 5.3 Mandibular remodeling after infancy. Remodeling is primarily confined to the ramus; the corpus is relatively stable.

information derived from *lateral* cephalograms. A completely different relationship between the amount of deposition and resorption may be indicated by a more comprehensive, three-dimensional analysis.

Condylar Cartilage Growth and Mandibular Rotation

In a detailed cross-sectional histological study of the condylar cartilage of rhesus monkeys at five different maturational levels, Carlson, McNamara, and Jaul (1978) suggested that the relative thickness of the various components of this cartilage are related to age-associated changes in the function and growth of the temporomandibular joint. They indicated that the absolute thickness of the fibrocartilagenous articular layer, which caps the condyle, increases somewhat with age in the superior and posterosuperior regions. However, when evaluated *relative to overall size of the condyle,* this layer exhibited a significant decrease in thickness, particularly in the posterosuperior and posterior regions. The overall contribution of the posterosuperior and superior regions to the total articular-layer thickness increases with age. The thickness of articular cartilage is generally acknowledged to be related to the amount of stress occurring at that joint surface. The observation of a shift in the thickest part of the articular-cartilage layer to a more superior location on the condyle (Carlson et al. 1978) coincides temporally with the present observation of the direction of condylar growth becoming more vertical. Also, the anterosuperior and superior regions of the condyle, which presumably withstand the greatest stress when compressed or sheared against the articular eminence during mastication, come to occupy more *posterior,* and presumably less loaded, positions along the condyle during growth (fig. 5.4).

According to Carlson and coworkers, the relative thickness of the growth layer of the condylar cartilage, which is comprised of prechondroblastic and chondroblastic layers, is very similar

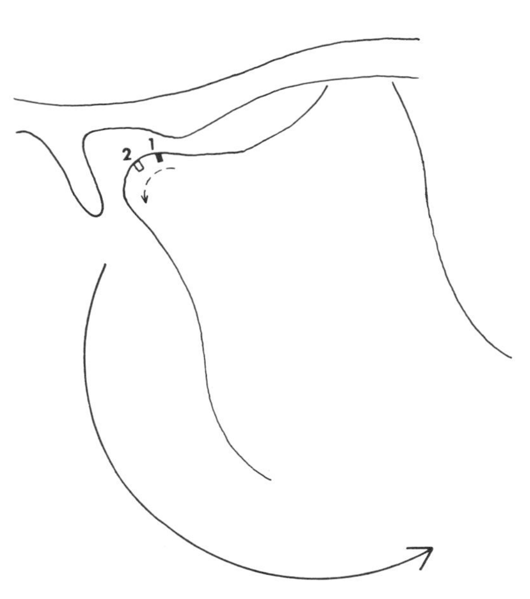

FIGURE 5.4 Schematic illustration showing the shift of the anterosuperior aspect of the condyle to a more posterior position with the counterclockwise rotation of the whole mandible that occurs with normal growth.

in neonates, infants, and juveniles and decreases in the older age groups. This finding is consistent with the present cephalometric evidence of greatest condylar growth in preadolescents, with a significant deceleration of growth at this site during adolescence. Carlson and associates also found that the posteriosuperior region was relatively and absolutely the thickest portion of the growth layer since birth. In the transition from neonate to infant, the increase in thickness in this region is profound (44.5%) and is consistent with the present observation of the highest rates of posterosuperior growth at condylion during infancy. This region attains a peak in thickness relative to the superior and posterior regions during the juvenile phase. After this phase, each of the regions decreases in thickness, becoming relatively evenly distributed during adulthood.

The present evidence of significant deceleration of the horizontal component of growth at condylion between 1.5 and 3 years of age concurs well with the histological picture of a shift in the thickest region of the cartilage from the posterior to a more superior position along the condyle during this period. In contrast to Carlson and coworker's report, however, it was found that maximal counterclockwise mandibular rotation occurs somewhat later than the juvenile period, at 3.5 or 4 years (adolescence) in males. Also during or slightly before this period a peak in the rate of anterior displacement of the mandible was observed. Thus, the shifting locus of condylar growth cartilage thickness seems temporally to precede the peak in mandibular rotation. This disparity in results between the present study and this previous histological study may be an artifact of a mixed-sex sample in the earlier study; peak values for mandibular rotation occur in the females somewhat earlier.

Measurements of change that reflect primary growth processes are presumably under tighter regulation and should exhibit less variability than those that reflect secondary or compensatory processes. If this is true, then the postulation that changes in the direction of condylar growth occur in *response* to rotational changes in jaw position (Björk and Skieller 1972; Carlson et al. 1978) is not clearly supported by the data discussed below.

Two cephalometric measures that represent change in mandibular position, *horizontal mandibular displacement* (XMNDSPL1) and *mandibular rotation* (MRLCRL1), demonstrated considerably more variability than the two orthogonal measures of growth at the condyle (XCO1 and YCO1). For example, the 95% simultaneous confidence intervals for the average growth velocity curves for horizontal mandibular displacement (XMNDSPL1) average about 40% greater than that for horizontal growth at condylion (XCO1) during early adolescence and are more than two times greater during and after adolescence. This consistent pattern in variability and the precedence of the histological condylar cartilage changes to the cephalometrically determined morphological changes indicate that the hypothesis that rotational changes in mandibular position are *secondary* to differential activity at the condyle cannot be ruled out. Measurement error is not likely to be the source of the present findings since greater variability was found in the measurements based solely on bone implant position, rather than in those based in part on anatomic landmarks, the latter of which are typically fraught with more error.

In general, however, the relationship between temporal and regional variability in condylar cartilage thickness and the amount and direction of overall mandibular growth (Carlson et al. 1978) is strongly supported by these data. Strictly morphological observations will not alone resolve questions of causality in condylar growth. However, a more thorough analysis of these data, in conjunction with information provided by organ culture studies of the sort begun by Copray and coworkers (1983, 1985) and Shimshoni and coworkers (1984) may provide greater insight into the underlying mechanisms of growth. In these in vitro studies, the proliferation of condylar cartilage was observed in response to the continuous application of compressive forces.

Gonial Angle and Masseter Muscle Growth

Carlson (1983) investigated masseter muscle growth in a longitudinal study of eight subadult female rhesus monkeys. This cephalometric study used gold muscle implants in conjunction with tantalum bone implants to chart the position of the masseteric insertion. Use of this approach made it possible to describe the positional change of the masseter relative to the cranium as well as the mandible. Carlson demonstrated that the masseter grows primarily longitudinally and maintains a relatively constant anteroposterior relationship to the cranium during ontogeny. When the position of the masseter was examined relative to the mandible, it was found to migrate posterosuperiorly. This migration was coordinated with the posterosuperior repositioning of the gonial region of the mandible. It is of interest to note that maximal rate of superior displacement of gonion (YGO1) occurs at 3.5 years in males, at which time horizontal growth at condylion peaks, as do the rates of anterior mandibular displacement and counterclockwise mandibular rotation. Collectively these findings might be interpreted to mean that all these aspects of mandibular growth occur in order to maintain a biomechanically constant configuration between the muscles of mastication and the tooth rows and temporomandibular joint, while permitting overall enlargement of all structures. In an experimental study involving altered mandibular posture in young adult monkeys, Carlson and Schneiderman (1983) found that the entire maxillary complex repositioned so as to minimize changes in the original length and position of the masseter muscle. This same sort of process, in which the underlying skeletal substrate adapts to preserve certain geometric relations that are biomechanically important, may be a general feature of normal growth.

Condylar Growth in the Rhesus Monkey Compared to the Human

Mandibular condyle growth in the rhesus monkey is consistently posterosuperior relative to the original occlusal plane with an emphasis on the vertical component. This contrasts with Björk's (1969) assessment of the average trend in the human as *anterosuperior*, using a similar radiopaque implant technique. This apparent species difference in the direction of condylar growth is, in part, attributable to differences in the choice of a horizontal plane of reference. In the present study, horizontal is defined as the original occlusal plane. In Björk's study, the sella-nasion line, which was used as the horizontal plane of reference, is at a substantial angle above the occlusal, palatal, and Frankfort horizontal planes in the human (Riolo et al. 1974) as well as in the rhesus monkey. Nevertheless, if Björk's subjects, which are presented as "representative" of the human population, are reoriented so that the occlusal plane is horizontal, species differences in the direction of condylar growth are still apparent, though less striking. Björk reported that the most common pattern of mandibular growth is one in which the mandible rotates counterclockwise; among these "forward" rotators, there is the category "Type II," in which the direction of condylar growth is strongly anterior. Extremes of this human type, in which the direction of condylar growth is *anterior* to the axis of the condylar neck (or ramus as a whole) were not observed in any of the monkeys in this study for any significant duration; there was no evidence of growth on the anterior aspect of the condyle in the rhesus monkey. Furthermore, in agreement with Elgoyen and coworkers (1972), the direction of condylar growth in the rhesus monkey appears less variable than in the human.

STATISTICAL ISSUES

Biological conclusions gained from data are completely dependent on the manner in which they are collected and analyzed. It has long been recognized that the quality of information regarding growth that can be extracted from longitudinal data is far superior to that available in cross-sectional data. Tanner (1962) states that twenty times more cross-sectional records than longitudinal records are required to estimate distance growth curves with the same accuracy. Tanner also states that cross-sectional data perform even worse for estimating average velocity curves, and are useless for determining the variability associated with such curves. Despite this widely acknowledged superiority of longitudinal data for studying growth, the few studies that have gone to great lengths to collect such data—for example, Riolo and coworkers (1974), Boersma and coworkers (1979), and Behrents (1985) on humans, and Sirianni and coworkers (1982) and Sirianni and Swindler (1985) on pigtailed macaques—have ironically used statistical methods designed and appropriate for cross-sectional data. The reason for this has been the lack of readily accessible statistical models for analyzing longitudinal data.

The current study is a first attempt at bridging this gap by applying suitable biostatistical models to a nontrivial longitudinal data set for a primate species. In such a preliminary effort, various problems arise. The following is a discussion of some of the contradictory or enigmatic findings that have emerged from applying these multivariate methods to data we are used to seeing described with conventional least-squares statistics.

Cohort Effects

A few of the growth velocity curves—for example, horizontal condylar growth (XCO1) and vertical growth at the mandibular alveolar ridge (YMNALVR1)—give the appearance of *cohort effects* (van't Hof et al. 1977), a potential risk of mixed longitudinal designs. Cohort effects are said to occur when different cohorts of subjects within a study exhibit discontinuous trends. The present study shows the appearance of such discontinuities at the junctions between the three segments of the growth velocity curves in the above variables. However, despite this appearance, this sort of effect is minimal since the male data set is more longitudinal than mixed, that is, the majority of animals are observed in all three segments.

Curve Fitting

The growth curve fitting method of Rao (1959) and Schneiderman and Kowalski (1985) can be misleading as to the direction a trend is taking when applied to short periods of time. This method, when used in conjunction with the method used for filling in missing data, tended to exaggerate the linearity of trends. This effect is pronounced at the beginning and end of certain growth curve segments (see, e.g., maxillary rotation relative to the CRL or MXRLCRL1), particularly when only four or fewer time points were considered. This problem was minimized, however, by examining data, even when scanty or incomplete, for the time points both before and after the fitted curve. In general, the examination of all available data, even when they could not be used in the multivariate analyses, tended to provide more cogent pictures of how particular variables behaved.

Kowalski and Guire (1974) made the very convincing argument that, when fitting a curve to longitudinal data, the goodness-of-fit criteria should be subordinate to choosing a function that "accurately mirrors the biological structure of the process under consideration" (p. 140). In cases where a polynomial growth curve generated using Rao's method does not make sense biologically, this particular mathematical representation of the data may be less than optimal.

Rao's Paradox

Rao (1966), Healy (1969), and Kowalski (1972) discussed the situation that has come to be known as *Rao's paradox*. In this situation, variables that differ individually between two groups (according to univariate *t*-tests) do not show differences when considered together in a vector using Hotelling's T^2 test. The explanation is that the inclusion into the test of variables that do not include information about group differences will diminish the ability of the multivariate test to demonstrate significance. This sort of result occurred in some instances in the present data set, but so did the converse. An example of Rao's paradox is found in the male data for horizontal maxillary displacement (XMXDSPL1) for the 6- to 7.5-year period. In this situation the individual velocities for each of the half-year intervals attain univariate significance at the 0.042, 0.011, and 0.13 levels, respectively, but the vector of velocities only attains significance at the 0.13 level.

For this same measure of maxillary displacement (XMXDSPL1), but over the 3- to 6-year period, the converse of Rao's paradox occurs. The vector of decelerations over this period demonstrates significance at the 0.0007 level, yet only one deceleration for a single half-year interval demonstrates univariate significance at the 0.05 level. Consideration of all of the available multivariate and univariate information allows us to make some sense of this apparent paradox. The narrow confidence band indicates a uniformity of behavior over time among the subjects that is not readily apparent from examination of the univariate standard deviations. Thus, strong decelerations for the 3.5- to 5.5-year observations, only the last of which attains univariate significance, emerge as significant when considered collectively and serially. Furthermore, the univariately significant deceleration between 5 and 5.5 years suggests that a reduction in the rate of the forward displacement of the maxilla may make the greatest contribution to the overall trend that spans 3 years.

In the craniofacial variables examined longitudinally in the study, the Hotelling's T^2 test (part of Hills's [1968] approach; see Schneiderman and Kowalski 1989) tended to demonstrate vectors of velocities and accelerations that were significantly different from zero less frequently than the univariate *t*-tests performed on the individual velocities and accelerations. In general, the multivariate test appears more conservative and stringent. When used in this context, Hotelling's T^2 test is appropriate, as it will tend to prevent individuals that exhibit extreme trends (over a number of observations) from biasing the results. The contribution of such individuals in univariate tests performed at isolated time points will not be moderated in this fashion, since the covariance structure of the data set is not considered. This use is perhaps a more justifiable application of multivariate tests (such as the T^2) than their more usual application to completely different variables measured at the same time.

On the one hand, multivariate tests are more conservative, as illustrated by the present results. On the other hand, as Kowalski (1972) has pointed out, the power and robustness of multivariate tests are virtually unknown. Despite the fact that multivariate tests are computational extensions of their simpler, univariate analogs, many (though not the ones used here) involve assumptions concerning the multivariate normal distribution, whose properties remain largely unknown.

6

CONCLUSIONS

In the *1979 Yearbook of Physical Anthropology*, Sirianni and Swindler made a strong case for the need for standards of craniofacial growth based on longitudinal data, generated using the most sophisticated multivariate statistical approaches available. They stressed the importance of using long-term longitudinal rather than mixed-longitudinal data, as only from the former can one construct accurate distance and velocity curves. Furthermore, the former type of design, when used in conjunction with the radiopaque bone-implant technique, permits the construction of a temporally refined picture of local changes in morphology due to remodeling (Sirianni and Swindler 1979). The application of a suitable multivariate polynomial growth curve fitting technique to good long-term male rhesus monkey data has made it possible to answer Sirianni and Swindler's call. Estimates of central tendency and variability for the rhesus monkey population for a number of measures of local and dimensional change in the craniofacial skeleton that can only be determined from serial longitudinal data were calculated in a manner that does not ignore the intrinsic properties of longitudinal data.

SUMMARY OF FINDINGS

The transition from the facial form of the infant to that of the adult involves the complex interplay of the enlargement, remodeling, and displacement of the various skeletal components as well as associated soft-tissue changes. The following is an effort to consolidate the detailed results of this study so as to provide a coherent overview of maxillomandibular growth in the male rhesus monkey.

The maxilla and mandible of the rhesus monkey enlarge and reposition rapidly throughout infancy and the juvenile period. Sometime between 3 and 4 years of age the rates of growth throughout the maxillomandibular complex generally begin to decelerate. This deceleration continues throughout adolescence and well past the age of reproductive maturity, up to perhaps 7 years of age in males. In marked contrast to the human, in the rhesus monkey the facial skeleton grows most rapidly during the earliest phases of life and the rate of growth generally declines thereafter. There is no pronounced preadolescent decline in growth, nor is there unequivocal evidence of major accelerations during adolescence, as occurs in the human.

In the course of its enlargement, the maxillary complex displaces in a predominantly forward direction relative to the cranial base; the downward component of this displacement is consistently less than half the forward component throughout ontogeny. The forward displacement of the maxilla coincides temporally with and is approximately equivalent metrically to the marked deposition of bone along the posterior aspect of the maxillary tuberosity. Over the first 4 years of life, the rate of horizontally directed maxillary displacement is strong and relatively constant; after this time it decelerates. During the 4- to 7-year age span, the amount of forward displacement is somewhat greater than the amount of bone growth at the tuberosity. Between 6.5 and 7.5 years of age, both the anterior displacement of the maxilla and posteriorly directed growth at the tuberosity cease.

The greatest contribution to the overall lengthening of the maxillary complex is provided by bone apposition at the maxillary tuberosity. Apposition of bone along the facial aspect of the maxilla, for example, below the nasal aperture, contributes 20%

to 30% of the full adult length. This is in contrast to the human situation, in which the comparable aspect of the maxilla is resorptive after early childhood (Enlow 1966, 1968). Due to the relative instability of bone implants in the premaxilla, it was not possible to ascribe a precise contribution to overall maxillary lengthening by growth at the premaxillary suture. However, the graphic information provided by the composite plots suggests that this contribution is not major. The premaxillary suture is situated so that it is more likely involved in the lateral expansion of the maxilla to accommodate the succedaneous incisors than the addition of sagittal depth to the maxilla as a whole.

Addition of bone at the superior and, to a lesser extent, the inferior aspects of the maxillary tuberosity is the primary source of increases in posterior maxillary height. In general, increments in the vertical dimension of the maxilla are modest compared to those occurring in the horizontal dimension. This observation accords well with the finding of the predominantly greater displacement in the horizontal dimension than the vertical.

Of interest is the finding that the pattern of growth over time of the facial aspect of the maxilla more closely resembles that of the maxillary alveolar ridge than that for the maxillary tuberosity. While the relatively low velocity curves for the former sites are smooth and gradual, the latter, stronger growth velocity curve is more episodic. Thus, the periosteal depository activity at the tuberosity would appear, at least from the temporal evidence, to be under different influences than the other, slower growing regions.

The present findings largely resolve the controversy regarding the direction of maxillary growth and repositioning. Consistent with the findings of McNamara (1972) and McNamara, Riolo, and Enlow (1976), horizontal growth and displacement of the maxilla was found to predominate over its vertical counterpart at all observations. An earlier interpretation, that vertical growth and translation predominates, which was based largely on cross-sectional dry-bone histology (Enlow 1968) and limited longitudinal cephalometric evidence (Gans and Sarnat 1951), can be discarded.

Enlow's (1966, 1968) qualitative descriptions of bone remodeling within the jaws are, for the most part, corroborated by the present cephalometric results. The picture is refined, however, by the observation that growth at the maxillary tuberosity is not purely posterior in direction. Bone is not just added to the posterior aspect of the upper part of the maxillary tuberosity, but is added to its posterosuperior aspect as well. Specifically, there is a *significant*, though not enormous, superiorly directed component to the overall displacement of the pterygomaxillary fissure inferior point relative to the body of the maxilla until about 5 years of age. Also, the circumpubertal peak in the superiorly directed deposition of bone at this site coincides temporally with a peak in the rate of inferior displacement of the maxilla as a whole.

Like the maxillary complex, the mandible also displaces forward and downward relative to the cranial base, with the forward component predominating. Despite the fact that the rate of forward mandibular displacement exceeds the downward component by 50% or 60%, the downward component is still notable, decelerating from about 5 mm per year during the juvenile period to about 3 mm per year during adolescence.

In contrast to most of the aspects of maxillomandibular growth examined in this study, the forward component of mandibular displacement exhibits a distinctly sigmoid velocity curve. That is, there is a peak in the rate of forward displacement at about 3.5 years, followed by a sharp deceleration with a nadir at 5.5 years, immediately followed by another peak in activity. This aspect of mandibular repositioning is most likely related to the emergence and eruption of the permanent dentition into occlusion. Specific hypotheses concerning the details of this relationship remain to be formulated and tested using the digitized coordinates collected in this study.

The mandibular condyle is unequivocally the primary site at

which mandibular lengthening occurs. The rate at which the condyle is lengthened is the greatest of any site examined in this study. The posterosuperiorly directed increments at condylion are approximately equivalent to overall increments in mandibular length at most observations. Overall enlargement of the mandible is primarily accomplished by remodeling of the ramal portion. Specifically, the posterior and anterior ramal borders become repositioned posteriorly and superiorly by apposition and resorption, respectively. Beyond infancy, the inferior and anterior borders of the mandibular corpus do not remodel markedly; the corpus is also reasonably constant. Addition of bone along the anterior aspect of the corpus accounts for only 10% to 20% of total increments in mandibular length, the balance being contributed by condylar/ramal growth.

The rates of bone growth at the major sites of activity, though topographically separated from one another, are highly correlated. Specifically, the posteriorly directed growth at the maxillary tuberosity, which occurs by periosteal apposition, is highly correlated and therefore temporally coordinated with the intense endochondral bone formation occurring at the condyle. Even more tightly coordinated is the anterior displacement of the mandible with that of the maxilla. The coordination in the vertical displacement between the two jaws is less pronounced than that prevailing in the horizontal direction. In general, coordination in jaw *displacement* is greater than the coordination occurring between *growth rates at any two local sites of remodeling*. This coordination in the development of the intermaxillary relationship, which ultimately relates to occlusal efficiency, is greater than correlations among the actual sites of growth. This finding can be interpretated in light of Petrovic and his associates' extensive experimentation on mandibular growth in the rat (Petrovic et al. 1981). They have concluded that the rate and amount of growth at the mandibular condyle will adjust in a compensatory fashion to bring about the "normal shape" of the mandible when various perturbations are introduced. Their cybernetic model postulates that the occlusion—specifically, the intercuspation of the teeth—functions as a *comparator* in a servosystem that regulates the overall growth of the mandible. Thus it appears as if there is some mechanism, of which neuromuscular feedback (proprioception) may be a major component, that regulates and coordinates the *variable* growth activities at various sites. That the highest correlations in the present study were found in those pairs of measurements that reflect this comparator, that is, the sagittal maxillomandibular relationship, lends indirect support to this interpretation. Needless to say, this is an area that requires much additional investigation.

The direction of growth at the condyle (approximately 45 degrees above the occlusal plane) over most of the period of active growth in conjunction with the relocation of the ramus relative to the corpus results in the mandible growing more in length than in height. That is, increments in oblique mandibular length are greater than those in ramal height. This horizontal emphasis in overall mandibular growth (as well as in displacement relative to the cranial base) results in the pronounced lower facial prognathism characteristic of cercopithecoid monkeys. This contrasts the human situation in which the vertical component of growth and displacement is more pronounced, leading to the characteristically more orthognathic configuration.

The posterior parts of the jaws do not displace forward and downward relative to the cranial base in exactly the same proportions as the anterior parts. The relative contribution of the downward component of the overall displacement is relatively greater toward the posterior of the mandible. The same is true in the maxilla, but to a lesser degree. Thus the jaws of the rhesus monkey rotate counterclockwise as they are translated anteroinferiorly. Adjustments in the occlusal plane were found to offset the bodily rotation of each of the jaws. The occlusal plane rotates clockwise relative to the mandible, thus maintaining the tooth row in a reasonably constant relationship with the cranium. Anatomically, this reorientation of the occlusal plane relative to the mandible is apparently due the differential vertical migration of the mandibular dentition, and it is greatest posteriorly (McNamara et al. 1976). The complementary clockwise shift in the occlusal plane relative to the maxilla would similarly appear to occur by the differentially greatest vertical dental migration occurring anteriorly. Establishing the details of these dental movements, and their precise temporal relationship to maxillomandibular growth, requires further investigation.

While the phenomenon of the counterclockwise rotation of the jaws and the associated shift in the occlusal plane has been qualitatively described (McNamara et al. 1976), the present report is the first *quantitative* demonstration of this phenomenon as it occurs throughout much of the life cycle. Of note is the finding that the direction of rotation is always counterclockwise and occurs at a significant rate until about 6.5 years of age in the male rhesus monkey. Also notable is the finding that the amount of rotation is greater in the mandible than in the maxilla. This is consistent with the findings of other studies on the rhesus monkey as well as on the human (Björk 1963).

It remains to be determined whether the intense growth at the maxillary tuberosity and mandibular condyle is the cause or result of the equally profound rotational and translational movements of the jaws relative to the cranium. One general area of investigation that may shed light on this problem concerns the biomechanical forces acting on these active sites of growth—the mandibular condyle and maxillary tuberosity. It is well demonstrated that the *experimental alteration* of mandibular position, and its effect on the biomechanics of the muscles of mastication, leads to alterations in the growth at the mandibular condyle and maxillary sutures in adult monkeys (Schneiderman and Carlson 1981, 1983, 1985; Carlson and Schneiderman 1983; Bosscher 1985) and in juveniles (Carlson, Ellis, et al. 1982). As the mandible and muscles of mastication become repositioned in conjunction with the enlargement of other soft tissues (e.g., tongue and airway), and dental eruption and migration, the direction and magnitude of biomechanical forces passing through the skeletal substrate probably change. On the basis of the studies cited above, it will be worthwhile to explore further the relationship between the pattern of normal growth at the maxillary tuberosity (and sutures) and mandibular condyle and the biomechanical forces passing through these and adjacent skeletal regions. The extent to which growth at the condyle and the maxillary tuberosity is intrinsically determined is extremely difficult to examine directly, but the role of extrinsic factors can be more readily addressed using a variety of techniques such as strain gage analysis and electromyography. Whether growth at the condyle and tuberosity is strictly secondary and compensatory to growth in the associated soft-tissues and spaces or whether the growth at these sites serves as a "pacemaker" for growth elsewhere in the craniofacial complex (see Carlson 1985, for discussion of this controversy) cannot be directly answered with this the sort of descriptive data presented in this study.

The sort of normative data presented here will make the results of numerous experimental studies of craniofacial growth in the rhesus monkey more informative and interpretable, so that experimental effects and natural variability can be distinguished. These data have already been used successfully as control data for assessing local adaptations at the maxillary tuberosity of experimental rhesus monkeys treated with bite-opening appliances (Bosscher 1985).

Although the primary sample of males used in this study is not large by epidemiological standards, it has permitted the production of the most fine-grained quantitative description of many aspects of craniofacial growth available to date for the rhesus monkey. Readers can make their own assessment of the significance of trends in this study, since the actual significance levels of both the univariate and multivariate tests are provided. Furthermore, the 95% confidence bands for the average growth curves are explicit probabilistic statements about true population parameters, given the present sample sizes and periods of time considered in the calculations.

LIMITATIONS OF THE PRESENT STUDY

From a methodological standpoint, various aspects of longitudinal data analysis require further development. A satisfactory method for dealing with missing data needs to be found. A method that uses simultaneously all of the observations available in a data set is sorely needed. The technique used in this study required rectangular blocks of data for the calculation of average growth curves and thus made poor use of all the data collected. Data from other individuals having incomplete records and data from the same individuals for earlier or later time points could not be used in these calculations due to the current limitations of this multivariate statistical model. Perhaps randomization tests (Zerbe and Walker 1977; Zerbe 1979a, 1979b, 1979c; Schneiderman et al., n.d.) and related procedures such as bootstrapping (Diaconis and Efron 1983) may lead to the development of methods that allow the maximal utilization of all the information present in a longitudinal data set. Randomization tests involve the comparison of statistics computed on the observed data as it is actually ordered with statistics computed for all the possible permutations of the data. This nonparametric technique offers enormous potential as it requires almost no assumptions regarding the distribution and structure of the data; however, it is computationally very "machine intensive." Similarly, bootstrapping uses high-speed computing to generate all possible subsamples of a data set, yielding information on its sampling properties, in order to arrive at better estimates of the true population parameters. Such flexible methods would have applications, as almost all major longitudinal data sets available for humans and other primates are highly irregular in structure, that is, there are seldom exactly T observations for all N individuals at all of the T time points considered in the study.

Multiple serial observations are required to address the most interesting questions regarding coordination or lack of coordination, at various sites of growth. We cannot accept the current methodological limitations in which all subjects must start and finish at the same time points in order to be included in a formal multivariate statistical analysis. Thus, the further development of techniques to analyze mixed-longitudinal data that consist primarily of long-term serial records is essential.

Statistical methods for formally testing hypotheses regarding longitudinal differences between and among groups must also be developed. Although the methods used in this study are a first step in this direction, they do not allow us definitively to answer questions regarding sexual dimorphism or growth differences between groups that exhibit different morphologies as adults. Development of such methods may permit the isolation of the local sources of aberrant growth that lead to the development of malocclusions. The presence of a small subsample of monkeys in the colony used in this study that exhibit mild malocclusions (e.g., anterior open bites) will be ideally suited to this sort of analysis.

Multivariate extensions of the statistical models used in this study that consider a number of different morphological variables simultaneously over time may also be useful in describing shape change as a function of maturation. However, it is important to note that theory concerning the multivariate normal distribution and the formal properties of statistical methods that deal with this distribution is largely unknown (Kowalski 1972). Perhaps finite element (Cheverud et al. 1983; Moss et al. 1985), tensor analysis (Bookstein 1983), and related methods (Siegel and Benson 1982; Olshan et al. 1982), when coupled with appropriate statistical approaches that effectively deal with intercorrelation, will provide optimal means for generating comprehensive (i.e., two- and three-dimensional) descriptions of shape change.

CAPTIVE VERSUS NATURAL POPULATIONS

One possible concern about the data presented in this study is that the observations have been made on a sample of captive laboratory monkeys. Clearly, this sample does not represent any specific natural breeding or geographic population. However, it must be noted that logistics and cost make it impossible to collect this sort of longitudinal data from wild populations. Additional investigations comparing absolute dimensions from this study with those taken on the full range of ages within a natural sample may provide insight as to the extent to which we can generalize from such laboratory samples. Such data are now available for comparison from the free-ranging population of rhesus monkeys at Cayo Santiago, Puerto Rico (King 1990).

A preliminary comparison between the CHGD monkeys and those from Cayo Santiago has been conducted recently (King and Schneiderman 1991). Five craniofacial dimensions were evaluated for seven age groups between 0.5 to 7 years of age in 141 Cayo monkeys and the present sample. Because the two data sets differ in structure, conservative conventional statistics were used for analysis. The laboratory monkeys were shown to be larger in cranial height and midfacial depth after three years of age. However, the majority of the tests (approximately 80%) showed no statistical differences between the samples. This study would suggest that craniofacial growth in laboratory rhesus monkeys is not fundamentally different from that in their free-ranging counterparts, at least not obviously so. Further investigation using several other samples of laboratory rhesus monkeys are, however, required to fully resolve this issue. Again, randomization tests (Zerbe and Walker 1977; Schneiderman et al., n.d.) may provide comparisons of these data sets that are statistically more appropriate than conventional, highly structured tests.

The origins and homogeneity of the various groups of monkeys studied also have a bearing on the extent to which we can apply findings from one to the other. The laboratory colony used in the present study originated from several different localities in South Asia. Many of the monkeys were wild-caught. Thus the results issuing from such a potentially genetically diverse sample are not likely to underestimate the true population variability for the species as a whole. On the other hand, this variability may tend to exceed that found in more recent laboratory colonies in which the the monkeys have been both bred and raised in captivity by commercial breeders. It would also tend to exceed that found in isolated, seminatural demes such as that at Cayo Santiago. This latter population has, in fact, been shown to be quite homogeneous in terms of craniofacial morphology (King 1990).

COMPREHENSIVE ANALYSIS OF LONGITUDINAL DATA SETS

The calculation of all the results presented in this study involved the use of dozens of computer programs and well over one hundred data files, making the overall process extremely cumbersome, time-consuming, and expensive. There is a major need for the sort of statistical approach used in this study to be implemented in the form of a "user-friendly" and comprehensive computer software system. Efforts are currently underway to develop such software using several high-speed programming languages for generating the code, in conjunction with a prototyping system for generating menu-driven user interfaces (see Schneiderman et al. 1990 and n.d.).

Standards of human craniofacial growth must be recalculated using this sort of approach. Longitudinal standards currently in use (e.g., Riolo et al. 1974; Broadbent et al. 1975; Boersma et al. 1979) report only conventional means and standard deviations. Clinicians and researchers currently using these standards have been given a spurious impression of low variability from these standards because of the use of inappropriate statistics. We have recently demonstrated that conventionally calculated confidence intervals for longitudinal data may underestimate the breadth of the true confidence band by as much as 50% (Schneiderman, King, and Kowalski 1991). The treatment objectives of orthodontists, maxillofacial surgeons, and other clinicians may have to be reevaluated when better estimates of the true variability in human populations are produced. Such improved standards are essential for having a sound basis for distinguishing treatment effects from natural variability.

Criteria are needed for sorting out the effects of growth from treatment. Additionally, an objective basis is needed for distinguishing among the effects of numerous treatment modalities currently in use for the correction of malocclusions and other craniofacial disharmonies. Among the major modalities of conventional edgewise orthodontics, functional appliance therapy, and orthognathic surgery, there are hundreds of specific techniques, none of which have been tested for efficacy using adequate standards of normal growth. The widespread use of longitudinal standards derived by conventional least-squares statistical methods has tended to exaggerate the significance of treatment effects.

The sort of approach used in the present study may also be useful in the development of methods for differentiating between subpopulations of humans and sorting the growth patterns of individuals into their appropriate natural categories. It may also provide the basis for assigning probabilities to outcomes of various therapies. If objective scientific methods are developed for predicting the growth of individuals, they will more likely issue from the rigorous statistical study of subpopulations, not individuals, despite this latter focus in the growth-curve literature.

SPECIFIC AREAS FOR FUTURE RESEARCH

Additional female rhesus monkeys for which good long-term data are available must be added to the sample to gain a better understanding of sexual dimorphism. Additional measurements from the comprehensive set of digitized coordinates collected for this study must be taken. The full array of standard cephalometric measurements (such as those reported in Riolo et al. 1974) should be taken to provide directly comparable values for the interpretation of other studies on the rhesus monkey as well as other primates. Also, a more comprehensive approach must be developed for describing dentitional development and maturity.

One related area that promises to be fruitful is the functional morphological analysis of maturation in the muscles of mastication. That is, investigation of the relationship between morphological (e.g., shape, position and size changes) and physiological aspects of muscle maturation, during normal and altered growth, may contribute to our understanding of the determination of variation in facial morphology. Specifically, comprehensive ontogenetic morphological data on the muscles of mastication at the gross level are needed to supplement the growing body of information on the muscles at lower levels of organization, for example, sarcomere and fiber length (Nemeth et al. 1983; Weijs and van der Wielen-Drent 1983; Carlson, Nemeth, and Dechow 1985) and on physiological attributes (Dechow and Carlson 1982a, 1982b; Dechow et al. 1983). By creating a comprehensive digitized data set, the first step has been taken in constructing an anatomical foundation for developing and testing hypotheses concerning the role of biomechanical changes in masticatory apparatus in craniofacial development and variability.

APPENDIXES

APPENDIX A
COMPUTER SOFTWARE AND HARDWARE

1. Data acquisition and preliminary manipulation
   1.1 *Hardware.* An integrated system consisting of a *Summagraphics Supergrid* digitizer with a serial RS232 interface with the *Zenith Z111-32* microcomputer (with a Winchester hard disk system) and a *ZVM 136* color monitor was used for digitizing the tracings of the radiographs. Plots of the digitized coordinate data were produced on the *Hewlett-Packard HP7475A* high resolution multicolor plotter, which was also interfaced to the microcomputer. A *Hayes Smartmodem 1200* was used for communicating with The University of Michigan's mainframe computer, an *Amdahl 5860*.
   1.2 *Storage media.* Data and programs were stored locally on the Winchester disk system as well as on floppy disks.
   1.3 *Software.* The digitizing program of the *Ceph-Master* software system by *Trilobyte Software, Inc.*, of Plymouth, Michigan, was used for digitizing the tracings on the above system. The graphic display programs of *Ceph-Master* were used for displaying the digitized skulls on the monitor, as well as to produce hard-copy records on the plotter. Because of its error checking protocol, The University of Michigan's *MCP/WINDOW* program was used to transfer the coordinate data to the mainframe computer reliably.

2. Data manipulation and statistical analysis
   2.1 *Hardware and operating systems.* The *Amdahl 5860* using the *MTS* (Michigan Terminal System) operating system was used for procedures requiring great speed and vast quantities of memory and work space. Some of the less complicated proce-

dures were done locally on the *Z111*, which uses the *Z-DOS* and *MS-DOS* operating systems.

2.2 *Software*

2.2.1 Data manipulation. Reformatting the coordinate data for subsequent analysis was done with the various utility programs written by the author with *MIDAS* (1976), the interactive statistical system provided by The University of Michigan's Statistical Research Laboratory.

2.2.2 Age estimation, etc. Various programs written in *ZBASIC* and run on the *Z111* were used to estimate chronological ages from the polynomial regression equations developed with *MIDAS* (see below). A *Pascal* program run on the *Amdahl* in conjunction with various *MIDAS* programs was used for calculating ages from the estimated birth dates and known film dates.

2.2.3 Calculating raw measurements. The *EXTRACT* program written by Richard Miller of the university's Center for Human Growth and Development was used for initially computing the raw dimensions and angles used in subsequent analysis.

2.2.4 Descriptive and conventional statistical analysis. *MIDAS* provided the relative time processing capability which permitted the calculation of differences/changes between successive films. Various types of regression analysis were conducted using *MIDAS* to estimate chronological ages where they were unknown.

2.2.5 Growth curve analysis. Programs specifically for longitudinal analysis were written using the *SAS* statistical system (SAS Institute Inc., Cary, N.C.), licensed by The University of Michigan, also operating on the *Amdahl*. This software system has a powerful matrix algebra programming language that permitted the implementation of otherwise obscure methods of longitudinal analysis (Schneiderman and Kowalski 1985).

APPENDIX B
AGE PREDICTION EQUATIONS

Linear, polynomial, and multiple regression equations suitable for predicting chronological age are presented below. All of the regressions presented are highly significantly different from zero ($p < 0.0005$). They are grouped according to gross maturational status and sex. Within each of these categories, they are ordered according to predictive accuracy. Those regression equations having the lowest standard errors and highest *R*-squared statistics were deemed the best.

The variable AGE.MOS is chronological age in months. The standard error (SE) of the regression, also expressed in months, is the square root of the error mean square. *R*-squared (R-SQR) expresses the proportion of the total variability in chronological age that can be explained by the morphological variables. N is the number of subjects used to formulate the equation. Note that for equations involving dental variables to be applicable, the teeth considered in the equation must be present in the subject. All of the other measurements are typically available for an individual of any age. Equations for both sexes treated together are provided for estimating age in subjects in which sex is unknown.

All Females

Eq. F4: AGE.MOS = – 90.09 + 3.77(PALLN) + RMHT(1.46)
(R–SQR = 0.813 SE = 14.83 N = 43)

Eq. F5: $AGE.MOS = 60.92 - 6.01(RMHT) + 0.17(RMHT^2)$
$(R\text{–}SQR = 0.796 \quad SE = 15.44 \quad N = 43)$

- $AGE.MOS = 29.60 - 2.06(MNLN) + 0.04(MNLN^2)$
  $(R\text{–}SQR = 0.778 \quad SE = 16.18 \quad N = 44)$

- $AGE.MOS = -24.36 + 0.13(PALLN) + 0.11(PALLN^2)$
  $(R\text{–}SQR = 0.776 \quad SE = 16.33 \quad N = 44)$

- $AGE.MOS = -40.49 + 0.80(MXLN) + 0.03(MXLN^2)$
  $(R\text{–}SQR = 0.757 \quad SE = 17.01 \quad N = 44)$

Immature Females

Eq. F1: $AGE.MOS = 12.34 - 5.74(LDM2ER) + 2.46(LM2RTLN)$
$(R\text{–}SQR = 0.944 \quad SE = 1.70 \quad N = 8)$

Eq. F2: $AGE.MOS = 22.77 - 0.35(UM2AN) + 2.90(UM2RTLN)$
$(R\text{–}SQR = 0.896 \quad SE = 2.31 \quad N = 8)$

Eq. F3: $AGE.MOS = 15.34 + 1.18(PALLN) - 2.07(UM2ER)$
$(R\text{–}SQR = 0.918 \quad SE = 2.35 \quad N = 10)$

- $AGE.MOS = 45.66 - 2.61(UM2ER)$
  $(R\text{–}SQR = 0.808 \quad SE = 3.36 \quad N = 10)$

- $AGE.MOS = 134.09 - 16.99(RMHT) + 0.70(RMHT^2) - 0.01(RMHT^3)$
  $(R\text{–}SQR = 0.832 \quad SE = 4.81 \quad N = 20)$

Adult Females
(permanent canines have emerged)

- $AGE.MOS = 92.14 - 17.18(UM3ER) + 2.92(UM3ER^2) - 0.17(UM3ER^3)$
  $(R\text{–}SQR = 0.663 \quad SE = 15.27 \quad N = 25)$

- $AGE.MOS = 90.643 - 5.26(UM3ER)$
  $(R\text{–}SQR = 0.511 \quad SE = 16.81 \quad N = 23)$

All Males

Eq. M1: AGE.MOS = – 4.84 – 0.48(MNLN) + 0.02($MNLN^2$)
(R–SQR = 0.941 SE = 8.24 N = 48)

- AGE.MOS = – 94.72 + 1.98 (MXHT) + 1.66(MNLN)
  (R–SQR = 0.937 SE = 8.48 N = 48)

- AGE.MOS = – 14.63 + 0.008(MXLN) + 0.03($MXLN^2$)
  (R–SQR = 0.918 SE = 9.65 N = 48)

- AGE.MOS = – 77.30 + 4.88(PALLN) – 0.007($PALLN^2$)
  (R–SQR = 0.848 SE = 13.15 N = 48)

- AGE.MOS = – 169.84 + 19.74(MXHT) -.33($MXHT^2$)
  (R–SQR = 0.843 SE = 13.38 N = 48)

- AGE.MOS = – 91.05 + 1.46(UPFHT) + 4.20(UPFDPTH) – CRBAN(0.74)
  (R–SQR = 0.834 SE = 13.92 N = 48)

- AGE.MOS = – 36.37 +0.34(RMHT) + 0.07($RMHT^2$)
  (R–SQR = 0.821 SE = 14.28 N = 48)

Immature Males

Eq. M2: AGE.MOS = – 44.026 – 1.71(UM2ER) + 2.88(UPFDPTH)
(R–SQR = 0.815 SE = 4.95 N = 13)

- AGE.MOS = 4.68 + 3.04(UM2RTLN)
  (R–SQR = 0.637 SE = 6.931 N = 12)
- AGE.MOS = 42.77 – 2.46(UM2ER)
  (R–SQR = 0.587 SE = 7.06 N = 13)

Adult Males

- AGE.MOS = 29.08 + 7.34(UM3RTLN) – 0.60(LM3AN) – 2.54(LM3RTLN)
  (R–SQR = 0.819 SE = 11.03 N = 34)

Both Sexes—Immature

- AGE.MOS = – 14.13 + 1.91(UPFDPTH) – 1.92(UM2ER)
  (R–SQR = 0.778 SE = 4.49 N = 23)
- AGE0.MOS = 43.37 – 2.41(UM2ER)
  (R–SQR = 0.637 SE = 5.61 N = 23)

APPENDIX C
STATISTICAL TABLES

Information for Tables 1–10, 13–26, 28, and 30–32

Statistics are presented for each change variable on separate pages, each page consisting of two tables. The upper tables provide statistics for each of the half-year time points considered separately. The lower tables present summary statistics for each of the three major periods of the entire time span considered in this study. The definitions of each of the items presented in each column, reading from left to right, are given for the upper and lower tables below. Probabilities (p) that individual values were equal to zero were determined using conventional *t*-tests. Probability values less than or approximately equal to 0.15 are presented in the tables. This presentation does not imply, however, that the values are statistically or biologically significant. The actual values are presented primarily for comparative purposes. Values greater than 0.15 are reported as NS (not significant). The p values listed in the tables are the *unadjusted*, calculated probabilities.

Bonferroni corrections (Alt 1982) are presented here to allow readers to decide for themselves whether a calculated probability that a change equals zero is significant. These extremely stringent corrections will protect against experimentwise error when a battery of univariate tests are used simultaneously. For sixteen tests considered simultaneously, a significance level of 0.0031 or less must be attained to maintain an overall 0.05 nominal significance level; for twelve tests, a significance level of 0.0042 is required; for four tests, 0.0125; and for three tests, 0.0167.

For this first set of tables, $\bar{x}$ refers to the mean *change* (i.e., *delta, difference, or increment*), not to an absolute dimension.

Note AGC and CI in columns 4 and 5 of the upper table, and the $p(\mathbf{v} = 0)$ and $p(\mathbf{a} = 0)$ in columns 6 and 9 of the lower table, as these are the most rigorous descriptors of the data available.

Upper-Table Columns

1. Estimated Age. Interval of estimated chronological ages in years. Values in this column are the upper limits of half-year intervals.
2. H.GRP. Half-year groupings of subjects.
3. N. Number of subjects in sample.
4. AGC. Average values that define the *average growth velocity curve.* These are the best unbiased estimates of the true growth velocity curve and are calculated using maximum likelihood methods as part of Rao's (1959) method (Schneiderman and Kowalski 1985). These are expressed in mm or degrees per half-year.
5. CI. Simultaneous 95% confidence intervals for the AGC calculated following Rao (1959). In mm or degrees.
6. $\bar{v}$. Average velocity for each of the half-year intervals. These first-order differences were calculated following Hills's (1968) method (Schneiderman and Kowalski 1989). In mm or degrees per half-year.
7. $p(\bar{v} = 0)$. Probability that $\bar{v}$ equals zero. These were determined using a conventional *t*-test.
8. $\bar{a}$. Average acceleration for each of the half-year intervals. These second-order differences were calculated following Hills's (1968) method (Schneiderman and Kowalski 1989). In mm or degrees per half-year squared.
9. $p(\bar{a} = 0)$. Probability that $\bar{a}$ equals zero [see criteria for significance in $p(\bar{v} = 0)$].
10. N.Ob. Number of observations. Total number of films for a particular H.GRP. These values are sometimes greater than the actual number of subjects since some subjects are represented by more than one film over a particular interval of time.
11. $\bar{x}$. Mean quantity of change calculated using least-squares method on N.Ob. In mm or degrees per half-year.
12. SD. Conventional standard deviation for N.Ob. In mm or degrees.
13. $p(\bar{x} = 0)$. Probability that $\bar{x}$ is equal to zero.

14–17. N.Ob., $\bar{x}$, SD, and p are presented for females for all variables.

Lower-Table Columns

1. Age Interval. Chronological age intervals in years (ages are estimated) for which Rao's (1959) polynomial growth curves were cal-

culated, and Hills's (1968) approach was applied (multivariate analyses).

2. N. Number of subjects.
3. T. Number of time points considered.
4. $\bar{v}$. Average velocity for the period spanned, expressed in mm or degrees per half-year.
5. $p(\bar{v} = 0)$. Probability that the average velocity is equal to zero.
6. $p(\mathbf{v} = 0)$. Probability that the vector of velocities for the period concerned is equal to zero. The significance level was generated using the multivariate Hotelling's T^2 test as part of Hills's approach.
7. $\bar{a}$. Average acceleration for the period concerned. Expressed in mm per half-year.
8. $p(\bar{a} = 0)$. Probability that the average acceleration is equal to zero.
9. $p(\mathbf{a} = 0)$. Probability that the vector of accelerations is equal to zero. From Hotelling's T^2.
10. tau_0. Intercept of polynomial equation, also known as alpha coefficient or 0th parameter. Note that this parameter pertains explicitly to the starting-time point indicated in the first column of this table, and not necessarily to the beginning of the shaded region corresponding to the growth velocity curve and confidence bands depicted in figures 4.1–4.19.
11. CI. 95% confidence interval for tau_0.
12. tau_1. Slope of polynomial equation if linear. Beta or first parameter of higher-order equations.
13. CI. 95% confidence interval for tau_1.

14–17. Higher-order parameters (i.e., tau_2 and tau_3) and confidence intervals (CI) are presented for a few variables for which they significantly improve the fit (tables 15, 23, 27).

Information for Tables 11, 13, 27, and 29

These tables report statistics for dimensional variables, that is, various *absolute dimensions* of the jaws. The format and symbolism used in these tables is similar to that of the first set of tables, but $\bar{x}$ is the actual mean rather than the change. Means were observed at the each of the estimated ages listed in the first column. The parameters tau_0, tau_1, etc., define the *average growth curve,* otherwise known as a *distance-traveled curve.* Also reported on these tables, in columns 9 and 13, are the 95% confidence intervals (conventional, univariate) for the dimensions. These are provided so that they can be compared with the more rigorously calculated simultaneous 95% confidence intervals in the fifth column. See number 14 above regarding parameters of orders higher than linear.

TABLE 1. Horizontal Displacement of the Maxilla per Half-Year, in mm (XMXDSPL1)

	Males												Females			
Est. Age (Yrs.)	H. GRP	N	AGC	±CI	$\bar{v}$	$p(\bar{v}=0)$	$\bar{a}$	$p(\bar{a}=0)$	No. Ob.	$\bar{x}$	SD	$p(\bar{x}=0)$	No. Ob.	$\bar{x}$	SD	$p(\bar{x}=0)$
0.5	1								1	3.688			1	4.232		
1.0	2								9	3.557	1.785	.0003	3	4.051		
1.5	3								15	2.545	1.384	.0000	4	3.853	0.904	.0034
2.0	4	14	2.307	0.709	2.319	.0000			24	2.422	0.810	.0000	10	2.530	0.820	.0000
2.5	5	14	2.321	0.466	2.470	.0000	0.151	NS	20	2.445	0.881	.0000	7	1.392	0.490	.0003
3.0	6	14	2.334	0.398	2.303	.0000	–0.166	NS	22	2.658	1.060	.0000	13	1.744	0.804	.0000
3.5	7	14	2.348	0.574	2.449	.0000	0.145	NS	18	2.886	0.707	.0000	10	1.690	0.753	.0001
4.0	8	13	2.341	0.408	2.490	.0000	–0.387	NS	17	2.328	0.494	.0000	14	1.287	0.846	.0001
4.5	9	13	1.927	0.284	1.972	.0000	–0.519	.112	15	1.673	0.700	.0000	13	1.239	0.622	.0000
5.0	10	13	1.514	0.271	1.487	.0004	–0.484	.148	12	1.413	0.735	.0000	8	0.780	0.511	.0035
5.5	11	13	1.100	0.382	0.837	.0000	–0.650	.047	13	0.844	0.537	.0001	6	0.681	0.598	.0384
6.0	12	13	0.686	0.546	0.811	.0001	–0.026	NS	12	0.814	0.590	.0006	12	0.509	1.176	NS
6.5	13	8	0.819	1.272	0.686	.0420	–0.588	.015	13	0.683	0.633	.0022	5	0.489	0.324	.0279
7.0	14	8	0.481	0.766	0.750	.0110	0.064	NS	7	0.710	0.791	.0551	1	0.796		
7.5	15	8	0.142	0.453	0.162	.1330	0.651	.116	9	0.396	0.741	.1475	1	0.334		
8.0	16								4	0.201	0.221	NS				
8.5	17								6	0.019	0.383	NS				
9.0	18								2	–0.177						
9.5	19								3	0.063	0.107					
10.0	20								3	0.058	0.113					

Summary Statistics (Males Only)

Age Interval (Years)	N	T	$\bar{v}$	$p(\bar{v}=0)$	$p(\mathbf{v}=0)$	$\bar{a}$	$p(\bar{a}=0)$	$p(\mathbf{a}=0)$	tau_0	±CI	tau_1	±CI
1.5–3.5	14	4	2.385	.0000	.0000	0.044	NS	NS	2.307	0.538	0.014	0.256
3.0–6.0	13	6	1.746	.0000	.0000	–0.413	.0002	.0007	2.755	0.475	–0.414	0.164
6.0–7.5	8	3	0.533	.0136	.1317	–0.262	.0771	.0560	0.820	0.867	–0.339	0.389

TABLE 2. Vertical Displacement of the Maxilla per Half–Year, in mm (YMXDSPL1)

	Males												Females			
Est. Age (Yrs.)	H. GRP	N	AGC	±CI	$\bar{v}$	$p(\bar{v}=0)$	$\bar{a}$	$p(\bar{a}=0)$	No. Ob.	$\bar{x}$	SD	$p(\bar{x}=0)$	No. Ob.	$\bar{x}$	SD	$p(\bar{x}=0)$
0.5	1								1	3.330			1	–0.224		
1.0	2								9	0.867	0.650	.0039	4	0.075	1.084	NS
1.5	3								15	0.723	0.817	.0041	4	0.589	0.406	.0624
2.0	4	14	1.059	0.343	1.058	.0000			24	0.582	0.809	.0018	10	0.745	0.640	.0050
2.5	5	14	0.827	0.301	0.860	.0005	–0.198	NS	20	0.419	0.658	.0104	7	0.720	0.694	.0336
3.0	6	14	0.595	0.442	0.870	.0134	–0.010	NS	22	0.792	0.919	.0006	13	0.441	0.604	.0219
3.5	7	13	0.769	0.437	0.723	.0052	–0.531	.1066	18	0.425	0.722	.0230	10	0.361	0.762	NS
4.0	8	13	0.603	0.403	0.511	.0437	–0.212	NS	17	0.421	0.776	.0397	14	0.249	0.564	.1224
4.5	9	13	0.436	0.390	0.066	NS	–0.445	NS	15	–0.025	0.767	NS	13	0.239	0.641	NS
5.0	10	13	0.270	0.399	0.301	.0582	0.235	NS	12	0.174	0.487	NS	8	0.225	0.598	NS
5.5	11	13	0.104	0.428	0.059	NS	–0.242	NS	13	0.211	0.664	NS	6	–0.258	0.178	.0164
6.0	12	13	–0.063	0.475	–0.165	NS	–0.224	.1400	12	–0.265	0.568	.1340	12	0.122	0.554	NS
6.5	13	8	–0.015	0.637	0.004	NS			13	–0.013	0.508	NS	5	0.095	0.609	NS
7.0	14	8	–0.107	0.309	–0.193	NS	0.041	NS	7	–0.162	0.689	NS				
7.5	15	8	–0.198	0.297	–0.152	NS	–0.197	NS	9	–0.183	0.356	NS				
8.0	16								4	–0.082	0.606	NS				
8.5	17								6	–0.063	1.216	NS				
9.0	18								2	0.495						
9.5	19								3	–0.193	0.370					
10.0	20								3	0.076	0.202					

Summary Statistics (Males Only)

Age Interval (Years)	N	T	$\bar{v}$	$p(\bar{v}=0)$	$p(v=0)$	$\bar{a}$	$p(\bar{a}=0)$	$p(a=0)$	tau_0	±CI	tau_1	±CI
1.5–3.5	14	4	0.782	.0000	.0001	–0.240	.0200	.0893	1.059	0.260	–0.232	0.195
3.0–6.0	13	6	0.249	.0518	.0101	–0.178	.0008	.0074	0.769	0.360	–0.166	0.077
6.0–7.5	8	3	–0.114	NS	NS	–0.078	NS	NS	–0.015	0.435	–0.091	0.265

TABLE 3. Change in Angle between Maxilla and Cranium per Half-Year, in Degrees (MXRLCRL1)

	Males												Females			
Est. Age (Yrs.)	H. GRP	N	AGC	±CI	$\bar{v}$	$p(\bar{v}=0)$	$\bar{a}$	$p(\bar{a}=0)$	No. Ob.	$\bar{x}$	SD	$p(\bar{x}=0)$	No. Ob.	$\bar{x}$	SD	$p(\bar{x}=0)$
0.5	1								1	2.015			1	–0.286		
1.0	2								9	0.526	2.465	NS	4	–1.691	2.559	NS
1.5	3								15	–0.333	1.605	NS	4	–1.213	0.837	.0626
2.0	4	14	–1.436	1.498	–1.407	.0194			24	–1.190	1.438	.0005	10	–1.138	1.785	.0745
2.5	5	14	–1.763	0.947	–1.805	.0012	–0.398	NS	20	–1.283	1.659	.0026	7	0.055	0.782	NS
3.0	6	14	–2.089	0.898	–2.082	.0405	–0.277	NS	22	–1.937	2.213	.0005	13	–0.940	1.698	.0691
3.5	7	14	–2.416	1.404	–2.402	.0002	–0.320	NS	18	–1.156	2.094	.0316	10	–0.664	0.992	.0633
4.0	8	13	–1.090	0.896	–1.394	.0400			17	–1.148	1.972	.0288	14	–0.090	1.097	NS
4.5	9	13	–0.966	0.653	–1.321	.0073	0.073	NS	15	–0.912	1.356	.0208	13	–0.621	0.843	.0209
5.0	10	13	–0.841	0.435	–0.699	.0544	0.622	NS	12	–0.367	0.926	NS	8	–0.222	1.055	NS
5.5	11	13	–0.717	0.302	–0.746	.0019	–0.047	NS	13	–0.732	0.744	.0040	6	0.062	0.536	NS
6.0	12	13	–0.592	0.363	–0.636	.0002	0.110	NS	12	–0.690	0.630	.0030	12	–0.307	1.168	NS
6.5	13	8	–0.698	1.496	–0.656	NS			13	–0.585	1.132	.0870	5	–0.656	0.626	.0791
7.0	14	8	–0.322	0.652	–0.369	NS	0.414	NS	7	–0.404	1.059	NS	1	–1.053		
7.5	15	8	0.054	0.528	0.045	NS	0.286	NS	9	–0.041	0.452	NS				
8.0	16								4	–0.501	0.295	.0427				
8.5	17								6	–0.063	1.216	NS				
9.0	18								2	0.603						
9.5	19								3	–0.661	0.506					
10.0	20								3	0.014	0.206					

Summary Statistics (Males Only)

Age Interval (Years)	N	T	$\bar{v}$	$p(\bar{v}=0)$	$p(\mathbf{v}=0)$	$\bar{a}$	$p(\bar{a}=0)$	$p(\mathbf{a}=0)$	tau_0	±CI	tau_1	±CI
1.5–3.5	14	4	–1.924	.0003	.0011	–0.332	NS	NS	–1.436	1.135	–0.327	0.600
3.5–6.0	13	5	–0.959	.0006	.0032	0.189	NS	NS	–1.090	0.705	0.125	0.207
6.0–7.5	8	3	–0.327	.0921	NS	0.350	NS	NS	–0.698	1.020	0.376	0.623

TABLE 4. Horizontal Displacement of the Maxillopalatine Junction Point per Half-Year, in mm (XMXPJ1)

		Males											Females			
Est. Age (Yrs.)	H. GRP	N	AGC	±CI	$\bar{v}$	$p(\bar{v}=0)$	$\bar{a}$	$p(\bar{a}=0)$	No. Ob.	$\bar{x}$	SD	$p(\bar{x}=0)$	No. Ob.	$\bar{x}$	SD	$p(\bar{x}=0)$
0.5	1								1	–2.400						
1.0	2								9	–3.004	2.280	.0042	3	4.000		
1.5	3								15	–2.108	1.604	.0002	4	–3.157	1.436	.0218
2.0	4								24	–2.100	1.432	.0000	10	–2.550	1.996	.0029
2.5	5	12	–1.978	1.033	–1.882	.0002			20	–2.038	1.656	.0000	7	–1.357	1.469	.0501
3.0	6	12	–2.002	0.636	–2.609	.0001	0.251	NS	22	–2.550	2.089	.0000	13	–1.574	1.541	.0031
3.5	7	12	–2.025	0.541	–1.899	.0000	–0.961	.1007	18	–2.314	1.689	.0000	10	–1.208	1.416	.0245
4.0	8	12	–2.048	0.856	–2.149	.0000	1.437	NS	17	–2.057	1.015	.0000	14	–1.204	1.549	.0122
4.5	9	12	–1.550	1.016	–1.480	.0010			15	–1.060	1.217	.0046	13	–1.227	1.272	.0046
5.0	10	12	–1.230	0.649	–1.260	.0009	0.220	NS	12	–1.096	0.923	.0017	8	–0.871	1.065	.0682
5.5	11	12	–0.910	0.378	–0.986	.0004	0.273	NS	13	–0.704	0.472	.0002	6	–0.606	0.766	.1105
6.0	12	12	–0.591	0.443	–0.557	.0037	0.429	.1050	12	–0.657	0.788	.0160	12	–0.603	0.911	.0427
6.5	13	7	–0.612	0.716	–0.601	.0445			13	–0.702	0.512	.0011	5	–0.482	0.230	.0094
7.0	14	7	–0.506	0.533	–0.612	.0645	0.257	NS	7	–0.416	0.591	.1120	1	–1.393		
7.5	15	7	–0.400	0.471	–0.355	NS	–0.010	NS	9	–0.288	0.286	NS	1	–0.549		
8.0	16								4	–0.161	0.371	.1262				
8.5	17								6	0.130	0.266	NS				
9.0	18								2	–0.190						
9.5	19								3	–0.093	0.296					
10.0	20								3	–0.038	0.171					

Summary Statistics (Males Only)

Age Interval (Years)	N	T	$\bar{v}$	$p(\bar{v}=0)$	$p(\mathbf{v}=0)$	$\bar{a}$	$p(\bar{a}=0)$	$p(\mathbf{a}=0)$	tau_0	±CI	tau_1	±CI
2.0–4.0	12	4	–2.135	.0000	.0000	–0.089	NS	NS	–1.978	0.780	–0.023	0.397
4.0–6.0	12	4	–1.071	.0000	.0017	0.307	.0346	NS	–1.550	0.767	0.320	0.308
6.0–7.5	7	3	–0.523	.0351	NS	0.123	NS	NS	–0.718	0.679	0.106	0.206

TABLE 5. Vertical Displacement of the Maxillopalatine Junction Point per Half-Year, in mm (YMXPJ1)

	Males												Females			
Est. Age (Yrs.)	H. GRP	N	AGC	±CI	$\bar{v}$	$p(\bar{v}=0)$	$\bar{a}$	$p(\bar{a}=0)$	No. Ob.	$\bar{x}$	SD	$p(\bar{x}=0)$	No. Ob.	$\bar{x}$	SD	$p(\bar{x}=0)$
0.5	1								1	0.190			1	2.020		
1.0	2								9	0.359	0.627	.124	4	–0.006	0.637	NS
1.5	3								15	–0.052	0.814	NS	4	–0.383	1.002	NS
2.0	4	14	–0.281	0.574	–0.329	.110			24	–0.266	0.624	.048	10	–0.247	1.270	NS
2.5	5	14	–0.395	0.517	–0.303	NS	0.026	NS	20	–0.063	0.761	NS	7	0.058	1.101	NS
3.0	6	14	–0.509	0.628	–0.617	.080	–0.314	NS	22	–0.618	0.906	.004	13	–0.215	0.963	NS
3.5	7	14	–0.623	0.844	–0.707	.035	–0.090	NS	18	–0.460	1.246	.136	10	–0.460	0.932	.1519
4.0	8	13	–0.670	0.807	–0.272	NS	0.488	NS	17	–0.270	1.473	NS	14	0.030	0.866	NS
4.5	9	13	–0.490	0.586	–0.215	NS	0.058	NS	15	–0.136	1.148	NS	13	–0.101	0.629	NS
5.0	10	13	–0.309	0.405	–0.220	NS	–0.006	NS	12	–0.159	0.601	NS	8	–0.090	0.581	NS
5.5	11	13	–0.129	0.337	–0.354	.105	–0.134	NS	13	–0.284	0.798	NS	6	0.322	0.669	NS
6.0	12	13	0.051	0.439	–0.070	NS	0.284	.036	12	–0.177	0.848	NS	12	–0.170	0.711	NS
6.5	13	7	–0.142	0.670	–0.260	NS			13	–0.144	0.577	NS	5	–0.244	0.502	NS
7.0	14	7	–0.006	0.303	0.159	NS	–0.079	NS	7	0.074	0.827	NS	1	0.077		
7.5	15	7	0.131	0.432	0.130	NS	0.419	NS	9	0.142	0.237	.111	1	0.025		
8.0	16								4	0.094	0.613	NS				
8.5	17								6	0.243	0.627	NS				
9.0	18								2	–0.086						
9.5	19								3	–0.304	0.493					
10.0	20								3	0.108	0.221					

Summary Statistics (Males Only)

Age Interval (Years)	N	T	$\bar{v}$	$p(\bar{v}=0)$	$p(\mathbf{v}=0)$	$\bar{a}$	$p(\bar{a}=0)$	$p(\mathbf{a}=0)$	tau_0	±CI	tau_1	±CI
1.5–3.5	14	4	–0.489	.021	NS	–0.126	NS	NS	–0.281	0.435	–0.114	0.233
3.0–6.0	13	6	–0.315	.055	NS	0.138	.116	NS	–0.850	0.861	0.180	0.210
6.0–7.5	7	3	0.009	NS	NS	0.195	NS	NS	–0.142	0.450	0.136	0.319

TABLE 6. Horizontal Displacement of the Pterygomaxillary Fissure Point per Half-Year, in mm (XPTMXFI1)

		Males											Females			
Est. Age (Yrs.)	H. GRP	N	AGC	±CI	$\bar{v}$	$p(\bar{v}=0)$	$\bar{a}$	$p(\bar{a}=0)$	No. Ob.	$\bar{x}$	SD	$p(\bar{x}=0)$	No. Ob.	$\bar{x}$	SD	$p(\bar{x}=0)$
0.5	1												1	−4.066		
1.0	2								9	−2.883	2.379	.0066	4	−4.430	3.001	.0599
1.5	3								15	−2.571	1.830	.0001	4	−2.539	0.552	.0027
2.0	4	14	−2.000	0.658	−1.924	.0000			24	−1.659	1.124	.0000	10	−2.206	1.261	.0004
2.5	5	14	−1.812	0.466	−1.755	.0000	0.168	NS	20	−1.942	1.405	.0000	7	−1.238	1.084	.0234
3.0	6	14	−1.622	0.355	−1.797	.0000	−0.041	NS	22	−1.792	1.567	.0000	13	−1.698	0.941	.0000
3.5	7	14	−1.433	0.398	−1.843	.0000	0.356	NS	18	−2.034	0.892	.0000	10	−1.552	0.661	.0000
4.0	8	13	−1.736	0.222	−1.440	.0000			17	−1.867	0.846	.0000	14	−0.976	1.096	.0054
4.5	9	13	−1.417	0.200	−1.226	.0000	0.616	.0514	15	−1.060	1.217	.0046	13	−0.978	0.683	.0002
5.0	10	13	−1.098	0.262	−0.980	.0058	0.246	NS	12	−0.981	1.370	.0306	8	−0.229	0.941	NS
5.5	11	13	−0.779	0.367	−0.796	.0000	0.184	NS	13	−0.837	0.488	.0000	6	−0.988	0.563	.0077
6.0	12	13	−0.460	0.489	−0.433	.0352	0.363	.0682	12	−0.461	0.748	.0562	12	−0.526	1.017	.1005
6.5	13	8	−0.404	0.643	−0.500	.0348	−0.217	NS	13	−0.518	0.582	.0075	5	−0.245	0.540	NS
7.0	14	8	−0.360	0.484	−0.186	NS	0.314	NS	7	−0.019	0.634	NS	1	0.087		
7.5	15	8	−0.316	0.536	−0.403	.0446	0.531	NS	9	−0.476	0.595	.0431	1	−0.275		
8.0	16								4	−0.143	0.494	NS				
8.5	17								6	−0.021	0.545	NS				
9.0	18								2	−0.176						
9.5	19								3	−0.003	0.404					
10.0	20								3	0.144	0.164					

Summary Statistics (Males Above, Females Below)

Age Interval (Years)	N	T	$\bar{v}$	$p(\bar{v}=0)$	$p(\mathbf{v}=0)$	$\bar{a}$	$p(\bar{a}=0)$	$p(\mathbf{a}=0)$	tau_0	±CI	tau_1	±CI
1.5–3.5	14	4	−1.729	.0000	.0000	0.161	.0988	NS	−2.000	0.499	0.189	0.189
3.5–6.0	13	5	−1.056	.0000	.0000	0.352	.0001	.0009	−1.736	0.175	0.319	0.108
6.0–7.5	8	3	−0.363	.0169	NS	0.048	NS	NS	−0.404	0.439	0.044	0.233
2.5–4.0	6	4	−1.318	.0002	.0093	0.070	NS	NS	−1.672	0.718	0.109	0.419
4.5–6.0	5	3	−0.681	.0008	.0523	0.190	NS	NS	−0.894	0.354	0.212	0.306

TABLE 7. Vertical Displacement of Pterygomaxillary Fissure Point per Half-Year, in mm (YPTMXFI1)

		Males											Females			
Est. Age (Yrs.)	H. GRP	N	AGC	±CI	$\bar{v}$	$p(\bar{v}=0)$	$\bar{a}$	$p(\bar{a}=0)$	No. Ob.	$\bar{x}$	SD	$p(\bar{x}=0)$	No. Ob.	$\bar{x}$	SD	$p(\bar{x}=0)$
0.5	1								1	4.222			1	−1.442		
1.0	2								9	0.342	1.463	NS	4	0.725	1.558	NS
1.5	3								15	0.562	3.043	NS	4	1.450	0.885	.0465
2.0	4	14	1.228	0.897	1.459	.0005			24	1.285	1.691	.0011	10	1.475	1.221	.0041
2.5	5	14	1.420	0.571	1.157	.0005	−0.301	NS	20	1.017	1.017	.0040	7	0.505	1.077	NS
3.0	6	14	1.613	0.545	1.626	.0009	0.469	NS	22	1.831	1.697	.0001	13	0.463	1.428	NS
3.5	7	13	1.130	1.517	1.520	.0003	−0.106	NS	18	1.207	1.792	.0109	10	1.160	1.013	.0056
4.0	8	13	0.924	1.199	1.127	.0112	−0.264	NS	17	1.361	1.586	.0027	14	0.095	1.321	NS
4.5	9	13	0.718	0.898	1.013	.0066	−0.114	NS	15	0.867	1.052	.0065	13	0.601	0.765	.0150
5.0	10	13	0.512	0.640	0.581	.0035	−0.432	NS	12	0.731	1.050	.0345	8	0.816	1.156	.0860
5.5	11	13	0.306	0.495	0.205	.1339	−0.376	.0647	13	0.003	1.087	NS	6	0.115	3.645	NS
6.0	12	13	0.099	0.561	0.378	.0131	0.173	NS	12	0.403	0.449	.0100	12	0.206	1.113	NS
6.5	13	8	0.324	0.699	0.379	NS			13	0.188	0.894	NS	5	0.176	0.390	NS
7.0	14	8	0.181	0.354	0.134	NS	−0.103	NS	7	0.198	0.888	NS	1	0.576		
7.5	15	8	0.039	0.713	0.031	NS	−0.245	NS	9	−0.078	0.885	NS	1	−0.012		
8.0	16								4	0.301	0.621	NS				
8.5	17								6	−0.057	1.319	NS				
9.0	18								2	−0.114						
9.5	19								3	0.622	0.367					
10.0	20								3	0.089	0.782					

Summary Statistics (Males Only)

Age Interval (Years)	N	T	$\bar{v}$	$p(\bar{v}=0)$	$p(\mathbf{v}=0)$	$\bar{a}$	$p(\bar{a}=0)$	$p(\mathbf{a}=0)$	tau_0	±CI	tau_1	±CI
1.5–3.5	14	4	1.441	.0000	.0001	0.020	NS	NS	1.228	0.680	0.193	0.358
3.0–6.0	13	6	0.782	.0005	.0733	−0.203	.0215	.0616	1.130	1.252	−0.206	0.282
6.0–7.5	8	3	0.181	.0814	NS	−0.174	NS	NS	0.324	0.477	−0.142	0.417

TABLE 8. Horizontal Displacement of Supradentale per Half-Year, in mm (XSD1)

Est. Age (Yrs.)	Males												Females			
	H. GRP	N	AGC	±CI	$\bar{v}$	$p(\bar{v}=0)$	$\bar{a}$	$p(\bar{a}=0)$	No. Ob.	$\bar{x}$	SD	$p(\bar{x}=0)$	No. Ob.	$\bar{x}$	SD	$p(\bar{x}=0)$
0.5	1								1	0.945			1	1.147		
1.0	2								9	0.568	0.767	.0571	4	0.509	1.460	NS
1.5	3								15	0.611	0.653	.0028	4	0.340	0.656	NS
2.0	4	14	1.197	0.440	1.034	.0000			24	0.981	0.736	.0000	10	0.735	0.669	.0070
2.5	5	14	1.001	0.359	1.358	.0000	0.323	NS	20	0.750	0.456	.0000	7	0.286	0.742	NS
3.0	6	14	0.805	0.482	0.829	.0022	−0.529	.0105	22	0.768	0.900	.0006	13	0.424	0.730	.0582
3.5	7	12	0.788	0.629	0.826	.0107	−0.065	NS	18	0.567	0.843	.0109	10	0.199	0.735	NS
4.0	8	12	0.718	0.452	0.697	.0284	−0.129	NS	17	0.655	1.237	.0442	14	0.141	0.549	NS
4.5	9	12	0.649	0.310	0.794	.0243	0.097	NS	15	0.689	0.989	.0173	13	0.104	0.498	NS
5.0	10	12	0.580	0.269	0.907	.0107	0.114	NS	12	0.920	0.997	.0085	8	0.687	0.395	NS
5.5	11	12	0.510	0.364	0.477	.0015	−0.430	.0981	13	0.367	0.433	.0100	6	0.200	0.328	NS
6.0	12	12	0.441	0.525	0.649	.0012	0.171	NS	12	0.745	0.553	.0007	12	0.420	0.548	.0223
6.5	13	7	0.379	0.572	0.279	.1372	−0.290	NS	13	0.461	0.478	.0046	5	0.385	0.414	.1062
7.0	14	7	0.327	0.371	0.475	.0349	0.196	NS	7	0.396	0.556	.1083	1	−0.216		
7.5	15	7	0.275	0.240	0.186	.1322	0.486	NS	9	0.168	0.538	NS	1	−0.476		
8.0	16								4	0.148	0.176	NS				
8.5	17								6	0.181	0.555	NS				
9.0	18								2	0.247						
9.5	19								3	0.139	0.145					
10.0	20								3	0.139	0.060					

Summary Statistics (Males Only)

Age Interval (Years)	N	T	$\bar{v}$	$p(\bar{v}=0)$	$p(\mathbf{v}=0)$	$\bar{a}$	$p(\bar{a}=0)$	$p(\mathbf{a}=0)$	tau_0	±CI	tau_1	±CI
1.5–3.5	14	4	0.996	.0000	.0000	−0.090	NS	.0568	1.197	0.333	−0.196	0.220
3.0–6.0	12	6	0.725	.0000	.0039	−0.035	NS	NS	0.788	0.500	−0.069	0.163
6.0–7.5	7	3	0.314	.0090	.0545	−0.047	NS	NS	0.379	0.384	−0.052	0.158

TABLE 9. Vertical Displacement of Supradental per Half-Year, in mm (YSD1)

Est. Age (Yrs.)	H. GRP	N	AGC	±CI	$\bar{v}$	$p(\bar{v}=0)$	$\bar{a}$	$p(\bar{a}=0)$	Males No. Ob.	Males $\bar{x}$	Males SD	Males $p(\bar{x}=0)$	Females No. Ob.	Females $\bar{x}$	Females SD	Females $p(\bar{x}=0)$
0.5	1								1	0.548			1	1.623		
1.0	2								9	0.185	0.837	NS	4	0.196	1.460	NS
1.5	3								15	0.200	1.062	NS	4	0.185	0.750	NS
2.0	4	14	0.581	0.570	0.618	.0028			24	0.629	0.694	.0002	10	0.607	0.960	.0765
2.5	5	14	0.358	0.278	–0.088	NS	–0.706	.1010	20	0.168	0.566	NS	7	–0.031	1.311	NS
3.0	6	14	0.134	0.427	–0.431	NS	–0.342	NS	22	–0.231	1.478	NS	13	0.260	1.038	NS
3.5	7	13	0.396	0.534	0.095	NS	0.698	NS	18	0.110	1.008	NS	10	0.191	0.746	NS
4.0	8	13	0.421	0.430	0.515	.0447	0.420	NS	17	0.335	0.921	NS	14	0.060	0.610	NS
4.5	9	13	0.446	0.345	0.660	.0059	0.145	NS	15	0.512	0.585	.0044	13	0.057	0.819	NS
5.0	10	13	0.470	0.294	0.440	.0992	–0.220	NS	12	0.330	0.817	NS	8	–0.163	0.585	NS
5.5	11	13	0.495	0.294	0.455	.0119	0.015	NS	13	0.422	0.844	.0964	6	0.129	0.601	NS
6.0	12	13	0.520	0.347	0.473	.0019	0.018	NS	12	0.502	0.504	.0055	12	0.109	0.520	NS
6.5	13	8	0.290	0.829	0.300	NS			13	0.282	0.621	.1270	5	0.541	0.616	.1211
7.0	14	8	0.150	0.276	0.138	NS	–0.122	NS	7	0.136	0.501	NS	1	0.595		
7.5	15	8	0.011	0.391	0.016	NS	–0.162	NS	9	0.040	0.363	NS	1	–0.548		
8.0	16								4	0.294	0.372	NS				
8.5	17								6	0.176	0.484	NS				
9.0	18								2	0.163						
9.5	19								3	0.237	0.494					
10.0	20								3	–0.376	0.224					

Summary Statistics (Males Only)

Age Interval (Years)	N	T	$\bar{v}$	$p(\bar{v}=0)$	$p(\mathbf{v}=0)$	$\bar{a}$	$p(\bar{a}=0)$	$p(\mathbf{a}=0)$	tau_0	±CI	tau_1	±CI
1.5–3.5	14	4	0.092	NS	.0190	–0.117	NS	NS	0.581	0.432	–0.224	0.318
3.0–6.0	13	6	0.440	.0012	.0205	0.076	NS	NS	0.396	0.440	0.025	0.106
6.0–7.5	8	3	0.151	.0713	NS	–0.142	NS	NS	0.290	0.565	–0.139	0.400

TABLE 10. Vertical Displacement of Maxillary Alveolar Ridge per Half-Year, in mm (YMXALVR1)

		Males											Females			
Est. Age (Yrs.)	H. GRP	N	AGC	±CI	$\bar{v}$	$p(\bar{v}=0)$	$\bar{a}$	$p(\bar{a}=0)$	No. Ob.	$\bar{x}$	SD	$p(\bar{x}=0)$	No. Ob.	$\bar{x}$	SD	$p(\bar{x}=0)$
0.5	1								1	1.186			1	1.715		
1.0	2								9	0.514	0.334	.0017	4	0.504	0.334	.0569
1.5	3								15	0.556	0.547	.0015	4	0.320	0.288	.1127
2.0	4	14	0.680	0.338	0.718	.0001			24	0.628	0.532	.0000	10	0.567	0.538	.0089
2.5	5	14	0.807	0.251	0.858	.0018	0.140	NS	20	0.519	0.365	.0000	7	0.173	0.541	NS
3.0	6	14	0.933	0.282	0.830	.0021	–0.028	NS	22	0.889	0.723	.0000	13	0.453	1.119	NS
3.5	7	13	0.547	0.690	–0.296	NS	0.258	NS	18	0.572	0.744	.0046	10	–0.023	0.804	NS
4.0	8	13	0.494	0.495	1.876	.1335	2.172	NS	17	0.579	1.189	.0619	14	0.382	0.648	.0459
4.5	9	13	0.441	0.332	0.550	.1493	–1.326	NS	15	0.377	1.119	NS	13	0.195	0.393	.0986
5.0	10	13	0.389	0.266	0.084	NS	–0.467	NS	12	0.178	0.519	NS	8	–0.180	0.713	NS
5.5	11	13	0.336	0.357	0.041	NS	–0.042	NS	13	–0.025	0.769	NS	6	0.338	0.793	NS
6.0	12	13	0.283	0.529	0.363	NS	0.321	NS	12	0.270	1.114	NS	12	–0.105	0.614	NS
6.5	13	8	0.376	0.615	0.325	.1392	–0.222	NS	13	0.267	0.613	.1430	5	0.454	0.330	.0370
7.0	14	8	0.233	0.399	0.291	.1262	–0.034	NS	7	0.276	0.517	NS	1	–0.079		
7.5	15	8	0.091	0.645	0.069	NS	0.187	NS	9	0.135	0.453	NS	1	0.729		
8.0	16								4	0.563	0.591	.1531				
8.5	17								6	0.248	0.252	.0610				
9.0	18								2	0.070						
9.5	19								3	–0.118	0.262					
10.0	20								3	–0.037	0.511					

Summary Statistics (Males Only)

Age Interval (Years)	N	T	$\bar{v}$	$p(\bar{v}=0)$	$p(\mathbf{v}=0)$	$\bar{a}$	$p(\bar{a}=0)$	$p(\mathbf{a}=0)$	tau_0	±CI	tau_1	±CI
1.5–3.5	14	4	0.873	.0000	.0000	0.123	.0505	NS	0.680	0.256	0.127	0.139
3.0–6.0	13	6	0.436	.0000	.0066	0.132	NS	.1208	0.547	0.570	–0.053	0.181
6.0–7.5	8	3	0.228	.0567	NS	–0.128	NS	NS	0.376	0.419	–0.143	0.328

TABLE 11. Maxillary Length, in mm (MXLN)

	Males								Females			
Est. Age (Yrs.)	H. GRP	N	AGC	±CI	No. Ob.	x̄	SD	±CI	No. Ob.	x̄	SD	±CI
0.5	1				6	25.600	5.521	5.794	4	24.394	0.760	1.210
1.0	2				13	29.118	3.080	1.861	5	29.406	1.662	2.064
1.5	3	12	32.155	2.136	23	32.877	2.958	1.280	8	30.419	4.336	3.625
2.0	4	12	35.127	1.613	26	36.818	3.087	1.247	11	35.052	3.657	2.457
2.5	5	12	38.098	1.210	21	38.834	2.422	1.103	8	39.679	1.104	0.923
3.0	6	12	41.070	1.073	23	42.151	3.100	1.341	13	40.754	2.217	1.340
3.5	7	12	44.041	1.291	18	45.786	2.807	1.396	13	41.314	2.593	1.568
4.0	8	13	48.495	2.529	17	48.598	3.308	1.701	16	43.652	2.955	1.575
4.5	9	13	50.018	2.328	15	50.988	2.888	1.599	13	44.900	1.514	0.915
5.0	10	13	51.540	2.210	12	53.841	2.646	1.681	9	47.135	1.900	1.460
5.5	11	13	53.063	2.189	13	54.529	3.095	1.871	9	47.306	1.558	1.198
6.0	12	13	54.586	2.267	12	55.828	2.951	1.875	12	48.114	1.502	0.954
6.5	13	8	57.598	3.411	13	56.764	2.600	1.572	5	48.808	2.120	2.633
7.0	14	8	58.200	3.267	7	58.028	2.262	2.092	1	46.722		
7.5	15	8	58.802	3.205	9	58.169	2.006	1.542	1	47.795		
8.0	16				4	59.094	3.213	5.113				
8.5	17				6	58.745	2.686	2.865				
9.0	18				2	61.493						
9.5	19				3	60.040	3.652	9.073				
10.0	20				3	59.747	3.952	9.818				

Summary Statistics (Males Only)

Age Interval (Years)	N	T	tau_0	±CI	tau_1	±CI
1.5–3.5	12	5	32.155	1.717	2.972	0.517
3.0–6.0	13	6	48.495	1.989	1.523	0.370
6.0–7.5	8	3	57.598	2.326	0.602	0.360

TABLE 12. Change in Maxillary Length per Half-Year, in mm (MXLN1)

Est. Age (Yrs.)	H. GRP	Males: N	AGC	±CI	$\bar{v}$	$p(\bar{v}=0)$	$\bar{a}$	$p(\bar{a}=0)$	No. Ob.	$\bar{x}$	SD	$p(\bar{x}=0)$	Females: No. Ob.	$\bar{x}$	SD	$p(\bar{x}=0)$
0.5	1								1	4.517			2	2.046		
1.0	2								9	3.912	2.177	.0007	4	7.373	3.144	.0184
1.5	3								15	3.065	1.557	.0000	6	2.712	3.107	.0855
2.0	4	12	3.163	0.744	3.176	.0000			24	3.536	1.928	.0000	10	3.693	1.754	.0001
2.5	5	12	3.025	0.587	2.849	.0000	–0.327	NS	20	3.200	1.593	.0000	7	1.653	1.306	.0155
3.0	6	12	2.886	0.610	3.164	.0000	0.315	NS	22	3.670	1.958	.0000	13	2.172	2.224	.0042
3.5	7	13	3.233	0.787	2.900	.0000	–0.264	NS	18	3.198	2.110	.0000	10	1.597	2.139	.0425
4.0	8	13	2.845	0.602	2.858	.0000	–0.573	NS	17	2.789	2.215	.0001	15	1.488	1.576	.0026
4.5	9	13	2.457	0.437	2.275	.0001	–0.584	NS	15	1.945	1.459	.0001	13	1.396	1.815	.0168
5.0	10	13	2.068	0.327	2.170	.0000	–0.104	NS	12	2.173	1.147	.0000	8	0.955	1.515	.1180
5.5	11	13	1.681	0.331	1.597	.0001	–0.574	.1168	13	1.114	0.661	.0001	6	1.066	0.976	.0441
6.0	12	13	1.293	0.446	1.294	.0000	–0.303	NS	12	1.297	0.641	.0000	12	1.360	1.592	.0130
6.5	13	8	0.999	1.192	1.000	.0127			13	1.197	0.838	.0002	5	1.018	0.893	.0635
7.0	14	8	0.762	0.735	0.744	.0217	–0.206	NS	7	0.747	0.738	.0366	1	1.015		
7.5	15	8	0.525	0.457	0.538	.0128	–0.256	NS	9	0.429	0.869	NS	1	1.162		
8.0	16								4	0.313	0.495	NS				
8.5	17								6	0.072	0.669	NS				
9.0	18								2	0.445						
9.5	19								3	0.169	0.348					
10.0	20								3	0.231	0.157					

Summary Statistics (Males Only)

Age Interval (Years)	N	T	$\bar{v}$	$p(\bar{v}=0)$	$p(\mathbf{v}=0)$	$\bar{a}$	$p(\bar{a}=0)$	$p(\mathbf{a}=0)$	tau_0	±CI	tau_1	±CI
1.5–3.5	12	4	3.022	.0000	.0000	–0.098	NS	NS	3.163	0.562	–0.139	0.260
3.5–6.0	13	6	2.271	.0000	.0000	–0.427	.0031	.0118	3.233	0.650	–0.388	0.172
6.0–7.5	8	3	0.760	.0047	.0513	–0.231	.1512	NS	0.999	0.813	–0.237	0.358

TABLE 13. Posterior Maxillary Height, in mm (MXHT)

	Males								Females			
Est. Age (Yrs.)	H. GRP	N	AGC	±CI	No. Ob.	x̄	SD	±CI	No. Ob.	x̄	SD	±CI
0.5	1				6	9.437	1.613	1.693	4	8.455	2.050	3.261
1.0	2				13	11.065	1.155	0.698	5	11.310	1.059	1.315
1.5	3	13	12.254	0.779	23	12.285	1.240	0.537	8	11.413	1.799	1.504
2.0	4	13	13.482	0.785	26	13.101	1.367	0.552	11	13.096	0.830	0.558
2.5	5	13	14.711	0.915	21	14.430	1.229	0.559	8	13.286	1.551	1.297
3.0	6	13	15.940	1.126	23	15.340	1.893	0.818	13	15.158	1.845	1.115
3.5	7	13	17.002	2.591	18	15.519	2.073	1.031	13	15.030	1.584	0.957
4.0	8	13	17.694	2.646	17	17.211	1.884	0.969	16	15.619	1.836	0.979
4.5	9	13	18.385	2.728	15	18.284	1.340	0.743	13	16.499	2.007	1.213
5.0	10	13	19.077	2.836	12	18.368	1.787	1.116	9	17.662	1.768	1.359
5.5	11	13	19.769	2.966	13	18.289	2.267	1.370	9	17.777	2.164	1.663
6.0	12	13	20.460	3.116	12	18.935	2.443	1.552	12	17.642	1.365	0.867
6.5	13	8	18.983	3.570	13	19.008	2.235	1.351	5	16.599	1.327	1.648
7.0	14	8	19.401	3.839	7	19.410	2.711	2.507	1	14.997		
7.5	15	8	19.820	4.202	9	19.395	2.670	2.053	1	15.517		
8.0	16				4	21.012	3.256	5.181				
8.5	17				6	21.614	2.768	2.905				
9.0	18				2	19.727						
9.5	19				3	20.550	0.165	0.411				
10.0	20				3	20.392	0.065	0.163				

Summary Statistics (Males Only)

Age Interval (Years)	N	T	tau_0	±CI	tau_1	±CI
1.5–3.5	14	5	12.254	0.634	1.229	0.264
3.5–6.0	13	7	16.311	2.095	0.692	0.228
6.0–7.5	8	4	18.564	2.463	0.419	0.489

TABLE 14. Change in Maxillary Height per Half-Year, in mm (MXHT1)

	Males												Females			
Est. Age (Yrs.)	H. GRP	N	AGC	±CI	$\bar{v}$	$p(\bar{v}=0)$	$\bar{a}$	$p(\bar{a}=0)$	No. Ob.	$\bar{x}$	SD	$p(\bar{x}=0)$	No. Ob.	$\bar{x}$	SD	$p(\bar{x}=0)$
0.5	1								2	2.251			2	5.485		
1.0	2								9	2.383	1.220	.0004	4	0.948	1.502	NS
1.5	3								15	0.112	1.999	NS	6	1.136	2.467	NS
2.0	4	13	1.434	0.588	1.077	.0025			24	0.867	1.728	.0219	10	1.547	1.805	.0240
2.5	5	13	1.249	0.322	1.539	.0000	0.462	NS	20	1.717	0.689	.0000	7	–0.005	3.014	NS
3.0	6	13	1.065	0.507	1.023	.0267	–0.516	NS	22	0.763	1.958	.0818	13	1.556	1.168	.0004
3.5	7	13	0.880	0.897	0.604	NS	–0.419	NS	18	0.326	3.136	NS	10	–0.215	1.581	NS
4.0	8	13	0.969	0.369	1.416	.0027	1.103	.1417	17	1.614	1.325	.0001	14	0.646	1.647	.1510
4.5	9	13	0.780	0.216	0.938	.0863	–0.478	NS	15	1.176	1.678	.0168	13	0.478	1.688	NS
5.0	10	13	0.591	0.223	0.546	NS	–0.393	NS	12	0.129	1.872	NS	8	0.782	1.057	.0748
5.5	11	13	0.403	0.380	–0.105	NS	–0.650	NS	13	–0.317	1.448	NS	6	–0.429	1.464	NS
6.0	12	13	0.214	0.576	0.268	NS	0.372	NS	12	0.315	1.456	NS	12	–0.128	2.155	NS
6.5	13	8	0.223	1.033	0.206	NS			13	0.254	0.912	NS	5	0.378	1.286	NS
7.0	14	8	0.388	0.609	0.395	.0616	0.144	NS	7	0.396	0.593	.1280	1	0.624		
7.5	15	8	0.553	0.832	0.539	.0961	0.189	NS	9	0.436	1.031	NS	1	0.563		
8.0	16								4	0.454	0.232	.0297				
8.5	17								6	0.329	1.050	NS				
9.0	18								2	–0.283						
9.5	19								3	0.666	0.489					
10.0	20								3	–0.101	0.185					

Summary Statistics (Males Only)

Age Interval (Years)	N	T	$\bar{v}$	$p(\bar{v}=0)$	$p(\mathbf{v}=0)$	$\bar{a}$	$p(\bar{a}=0)$	$p(\mathbf{a}=0)$	tau_0	±CI	tau_1	±CI
1.5–3.5	13	4	1.061	.0000	.0000	–0.158	NS	NS	1.434	0.453	–0.185	0.343
3.0–6.0	13	6	0.563	.0000	.0002	–0.009	NS	.1080	1.157	0.465	–0.189	0.177
6.0–7.5	8	3	0.380	.0829	NS	0.167	NS	NS	0.223	0.704	0.165	0.486

TABLE 15. Horizontal Displacement of the Mandible per Half-Year, in mm (XMNDSPL1)

	Males													Females			
Est. Age (Yrs.)	H. GRP	N	AGC	±CI	$\bar{v}$	$p(\bar{v}=0)$	$\bar{a}$	$p(\bar{a}=0)$	No. Ob.	$\bar{x}$	SD	$p(\bar{x}=0)$	No. Ob.	$\bar{x}$	SD	$p(\bar{x}=0)$
0.5	1								1	2.466			1	2.503		
1.0	2								9	3.808	1.605	.0001	4	1.872	1.252	.0582
1.5	3								15	3.316	2.676	.0003	4	3.740	1.566	.0059
2.0	4	14	3.076	1.079	2.950	.0000			24	3.149	1.621	.0000	10	2.492	1.993	.0033
2.5	5	14	3.154	0.770	3.717	.0000	0.767	NS	20	3.438	1.621	.0000	7	1.390	0.738	.0025
3.0	6	14	3.231	0.721	3.082	.0000	–0.636	NS	22	3.655	1.621	.0000	13	2.185	0.996	.0000
3.5	7	14	3.308	0.971	3.297	.0000	0.215	NS	18	3.504	1.461	.0000	10	1.945	1.022	.0002
4.0	8	13	3.304	1.979	3.289	.0000			17	3.306	0.923	.0000	14	1.681	1.840	.0046
4.5	9	13	2.821	1.549	3.009	.0000	–0.281	NS	15	2.543	1.032	.0000	13	1.520	1.233	.0008
5.0	10	13	1.818	1.169	1.677	.0000	–1.332	.0001	12	1.744	0.926	.0000	8	0.701	0.684	.0229
5.5	11	13	1.105	0.990	1.292	.0000	–0.385	NS	13	1.185	0.608	.0000	6	1.407	1.367	.0532
6.0	12	13	1.489	1.006	1.382	.0000	0.090	NS	12	1.510	0.748	.0000	12	0.260	1.369	NS
6.5	13	8	1.036	1.098	1.035	.0074			13	1.166	0.735	.0001	5	–0.086	2.028	NS
7.0	14	8	0.926	0.798	0.943	.0140	–0.130	NS	7	1.095	0.677	.0052	1	0.871		
7.5	15	8	0.816	0.911	0.814	.0100	–0.091	NS	9	0.981	1.037	.0219	1	0.910		
8.0	16								4	0.590	0.445	.0769				
8.5	17								6	0.097	0.827	NS				
9.0	18								2	–0.304						
9.5	19								3	0.145	0.082	0.093				
10.0	20								3	0.085	0.468	NS				

Summary Statistics (Males Only)

Age Interval (Years)	N	T	$\bar{v}$	$p(\bar{v}=0)$	$p(\mathbf{v}=0)$	$\bar{a}$	$p(\bar{a}=0)$	$p(\mathbf{a}=0)$	tau_0	±CI	tau_1	±CI	tau_2	±CI	tau_3	±CI
1.5–3.5	14	4	3.261	.0000	.0000	0.115	NS	NS	3.076	0.818	0.077	0.378				
3.5–6.0	13	5	2.130	.0000	.0000	–0.477	.0002	.0015	3.304	0.855	0.046	1.615	–0.664	0.867	0.135	0.125
6.0–7.5	8	3	0.930	.0029	.0569	–0.110	NS	NS	1.036	0.749	–0.110	0.421				

TABLE 16. Vertical Displacement of the Mandible per Half-Year, in mm (YMNDSPL1)

	Males												Females			
Est. Age (Yrs.)	H. GRP	N	AGC	±CI	$\bar{v}$	$p(\bar{v}=0)$	$\bar{a}$	$p(\bar{a}=0)$	No. Ob.	$\bar{x}$	SD	$p(\bar{x}=0)$	No. Ob.	$\bar{x}$	SD	$p(\bar{x}=0)$
0.5	1												1	2.446		
1.0	2								9	3.210	1.105	.0000				
1.5	3								15	2.587	3.452	.0116	5	2.977	0.888	.0017
2.0	4	14	2.449	0.486	2.479	.0000			24	2.008	1.934	.0000	10	3.321	2.434	.0019
2.5	5	14	2.218	0.365	1.890	.0010	–0.588	NS	20	1.935	2.002	.0004	7	2.037	0.661	.0002
3.0	6	14	1.988	0.576	2.013	.0002	0.123	NS	22	2.060	2.067	.0001	13	1.356	1.855	.0217
3.5	7	14	1.757	0.912	1.747	.0000	–0.265	NS	18	2.000	1.710	.0001	10	1.362	1.400	.0132
4.0	8	13	1.895	0.796	1.796	.0006	–0.532	NS	17	1.550	1.355	.0002	14	0.522	1.869	NS
4.5	9	13	1.447	0.669	0.858	.0738	–0.938	.0769	15	0.745	1.822	.1354	13	0.691	1.540	.1319
5.0	10	13	0.999	0.586	0.553	.0452	–0.304	NS	12	0.492	1.015	.1215	8	0.162	1.692	NS
5.5	11	13	0.551	0.567	0.332	.1247	–0.222	NS	13	0.421	0.818	.0884	6	–0.493	0.888	NS
6.0	12	13	0.103	0.618	0.160	NS	–0.172	NS	12	0.004	0.825	NS	12	0.840	2.045	NS
6.5	13	8	0.727	1.658	0.797	.1085			13	0.619	1.446	.1484	5	0.432	1.933	NS
7.0	14	8	0.166	0.637	–0.015	NS	–0.217	NS	7	–0.341	1.751	NS	1	–0.812		
7.5	15	8	–0.395	1.216	–0.232	NS	–0.812	NS	9	–0.228	1.309	NS	1	–0.394		
8.0	16								4	0.033	0.510	NS				
8.5	17								6	–0.028	0.852	NS				
9.0	18								2	0.670						
9.5	19								3	–0.240	0.654					
10.0	20								3	–0.011	0.311					

Summary Statistics (Males Only)

Age Interval (Years)	N	T	$\bar{v}$	$p(\bar{v}=0)$	$p(\mathbf{v}=0)$	$\bar{a}$	$p(\bar{a}=0)$	$p(\mathbf{a}=0)$	tau_0	±CI	tau_1	±CI
1.5–3.5	14	4	2.032	.0000	.0000	–0.244	.0856	NS	2.449	0.368	–0.231	0.294
3.0–6.0	13	6	1.004	.0004	.0017	–0.434	.0000	.0008	2.343	0.785	–0.448	0.168
6.0–7.5	8	3	0.183	NS	NS	–0.515	NS	NS	0.727	1.131	–0.561	0.892

TABLE 17. Change in Angle between Mandible and Cranium per Half-Year, in Degrees (MRLCRL1)

	Males												Females			
Est. Age (Yrs.)	H. GRP	N	AGC	±CI	$\bar{v}$	$p(\bar{v}=0)$	$\bar{a}$	$p(\bar{a}=0)$	No. Ob.	$\bar{x}$	SD	$p(\bar{x}=0)$	No. Ob.	$\bar{x}$	SD	$p(\bar{x}=0)$
0.5	1															
1.0	2								9	–1.864	1.366	.0035	4	–1.703	2.691	NS
1.5	3								15	–1.494	2.883	.0645	5	–3.417	4.162	.1402
2.0	4	12	–1.847	0.745	–1.816	.0028			24	–1.531	1.588	.0001	10	–1.244	3.410	NS
2.5	5	12	–2.012	0.762	–2.556	.0002	–0.740	NS	20	–1.826	1.760	.0002	7	–0.636	1.061	NS
3.0	6	12	–2.177	1.034	–2.189	.0015	0.367	NS	22	–1.601	1.534	.0001	13	–1.467	1.601	.0063
3.5	7	12	–2.342	1.423	–2.265	.0001	–0.076	NS	18	–1.762	1.369	.0000	10	–0.565	0.739	.0387
4.0	8	13	–2.210	1.537	–1.766	.0050			17	–1.634	1.611	.0007	14	–0.859	0.955	.0051
4.5	9	13	–1.800	1.263	–1.842	.0000	–0.076	NS	15	–1.374	0.993	.0001	13	–0.773	0.958	.0131
5.0	10	13	–1.389	1.002	–1.071	.0110	0.771	.0177	12	–0.937	1.259	.0257	8	–0.471	1.338	NS
5.5	11	13	–0.979	0.768	–0.952	.0016	0.118	NS	13	–0.899	0.836	.0022	6	–0.398	0.543	.1328
6.0	12	13	–0.568	0.595	–0.955	.0002	–0.002	NS	12	–0.939	0.786	.0016	12	–0.230	1.264	NS
6.5	13	8	–0.303	1.003	–0.247	NS	–0.234		13	–0.541	1.140	.1130	5	–0.149	0.709	NS
7.0	14	8	–0.642	0.556	–0.706	.0815	–0.459	NS	7	–1.071	1.013	.0312	1	–1.001		
7.5	15	8	–0.981	0.907	–0.940	.0187	–0.225	NS	9	–0.553	1.118	NS	1	–0.169		
8.0	16								4	–0.470	0.267	.0524				
8.5	17								6	0.125	1.157	NS				
9.0	18								2	0.489						
9.5	19								3	–0.399	0.405					
10.0	20								3	–0.111	0.567					

Summary Statistics (Males Only)

Age Interval (Years)	N	T	$\bar{v}$	$p(\bar{v}=0)$	$p(\mathbf{v}=0)$	$\bar{a}$	$p(\bar{a}=0)$	$p(\mathbf{a}=0)$	tau_0	±CI	tau_1	±CI
1.5–3.5	12	4	–2.206	.0000	.0005	–0.150	NS	NS	–1.847	0.563	–0.165	0.364
3.0–6.0	13	5	–1.317	.0001	.0088	0.203	.1511	.0032	–2.210	1.209	0.410	0.234
6.0–7.5	8	3	–0.631	.0041	.0387	–0.346	.1256	NS	–0.303	0.684	–0.339	0.530

TABLE 18. Horizontal Displacement of Condylion per Half-Year, in mm (XCO1)

	Males												Females			
Est. Age (Yrs.)	H. GRP	N	AGC	±CI	$\bar{v}$	$p(\bar{v}=0)$	$\bar{a}$	$p(\bar{a}=0)$	No. Ob.	$\bar{x}$	SD	$p(\bar{x}=0)$	No. Ob.	$\bar{x}$	SD	$p(\bar{x}=0)$
0.5	1												1	2.257		
1.0	2								9	4.533	1.963	.0001	4	1.436	1.135	.0853
1.5	3								15	3.346	1.421	.0000	5	4.076	1.890	.0085
2.0	4	14	3.443	0.799	3.339	.0000			24	3.256	1.173	.0000	10	2.780	1.693	.0006
2.5	5	14	3.046	0.595	3.275	.0000	–0.063	NS	20	2.839	1.598	.0000	7	1.644	0.671	.0006
3.0	6	14	2.650	0.509	2.585	.0000	–0.690	.0590	22	2.998	1.774	.0000	13	2.008	1.308	.0001
3.5	7	14	2.254	0.593	2.558	.0000	–0.027	NS	18	2.851	1.131	.0000	10	1.873	1.366	.0019
4.0	8	13	2.483	0.856	2.739	.0000	–0.406	NS	17	2.419	1.232	.0000	14	0.899	1.600	.0555
4.5	9	13	2.005	0.654	1.991	.0000	–0.748	.1129	15	1.629	1.122	.0001	13	1.050	1.178	.0074
5.0	10	13	1.527	0.476	1.408	.0001	–0.583	.1420	12	1.401	1.314	.0035	8	0.548	1.029	NS
5.5	11	13	1.049	0.359	0.935	.0017	–0.472	NS	13	0.482	1.244	NS	6	0.790	1.020	.1160
6.0	12	13	0.571	0.367	0.736	.0000	–0.200	NS	12	0.967	0.920	.0039	12	0.703	1.174	.0622
6.5	13	8	0.704	1.002	1.139	.0079			13	0.930	1.089	.0096	5	–0.115	1.873	NS
7.0	14	8	0.293	0.646	0.276	.0778	–0.022	NS	7	0.047	1.045	NS	1	–0.061		
7.5	15	8	–0.119	0.805	0.298	NS	–0.864	.0162	9	0.503	1.768	NS				
8.0	16								4	0.536	0.441	.0933				
8.5	17								6	0.101	0.954	NS				
9.0	18								2	0.006						
9.5	19								3	0.247	0.567	NS				
10.0	20								3	–0.052	0.156	NS				

Summary Statistics (Males Only)

Age Interval (Years)	N	T	$\bar{v}$	$p(\bar{v}=0)$	$p(\mathbf{v}=0)$	$\bar{a}$	$p(\bar{a}=0)$	$p(\mathbf{a}=0)$	tau_0	±CI	tau_1	±CI
1.5–3.5	14	4	2.939	.0000	.0000	–0.260	.1217	.0121	3.443	0.605	–0.396	0.232
3.0–6.0	13	6	1.826	.0000	.0005	–0.482	.0003	.0016	2.961	0.882	–0.478	0.188
6.0–7.5	8	3	0.571	.0241	.0943	–0.421	.0152	.0409	0.704	0.683	–0.411	0.436

TABLE 19. Vertical Displacement of Condylion per Half-Year, in mm (YCO1)

		Males											Females			
Est. Age (Yrs.)	H. GRP	N	AGC	±CI	$\bar{v}$	$p(\bar{v}=0)$	$\bar{a}$	$p(\bar{a}=0)$	No. Ob.	$\bar{x}$	SD	$p(\bar{x}=0)$	No. Ob.	$\bar{x}$	SD	$p(\bar{x}=0)$
0.5	1															
1.0	2								9	4.910	2.415	.0003				
1.5	3								15	4.678	2.247	.0000	5	6.210	2.158	.0030
2.0	4	14	3.866	0.585	3.803	.0000			24	3.869	1.975	.0000	10	4.850	2.368	.0001
2.5	5	14	3.896	0.527	4.044	.0000	0.241	NS	20	4.124	1.797	.0000	7	3.472	2.091	.0046
3.0	6	14	3.926	0.723	3.866	.0000	–0.178	NS	22	3.865	1.380	.0000	13	2.970	1.124	.0008
3.5	7	14	3.955	1.038	3.941	.0000	0.076	NS	18	4.432	1.090	.0000	10	2.312	0.952	.0000
4.0	8	13	3.526	0.833	3.828	.0000	–0.476	NS	17	3.665	1.282	.0000	14	0.899	1.600	.0555
4.5	9	13	2.919	0.691	3.454	.0000	–0.374	NS	15	3.107	1.694	.0000	13	1.744	1.058	.0001
5.0	10	13	2.312	0.627	1.919	.0000	–1.536	.0032	12	1.854	1.011	.0001	8	1.035	1.359	.0682
5.5	11	13	1.705	0.664	1.876	.0003	–0.043	NS	13	1.874	1.447	.0005	6	0.190	0.627	NS
6.0	12	13	1.098	0.788	1.422	.0001	–0.454	NS	12	1.361	0.890	.0003	12	0.432	0.858	.1092
6.5	13	8	1.024	1.206	1.056	.0064			13	1.096	0.815	.0004	5	1.364	1.983	NS
7.0	14	8	0.820	0.597	1.354	.0174	–0.866	NS	7	1.638	1.074	.0068	1	1.141		
7.5	15	8	0.616	0.509	0.488	.0152	0.299	NS	9	0.635	1.675	NS	1	–0.067		
8.0	16								4	0.173	0.457	NS				
8.5	17								6	0.131	0.591	NS				
9.0	18								2	0.154						
9.5	19								3	0.048	0.569					
10.0	20								3	0.131	0.898					

Summary Statistics (Males Only)

Age Interval (Years)	N	T	$\bar{v}$	$p(\bar{v}=0)$	$p(\mathbf{v}=0)$	$\bar{a}$	$p(\bar{a}=0)$	$p(\mathbf{a}=0)$	tau_0	±CI	tau_1	±CI
1.5–3.5	14	4	3.913	.0000	.0000	0.046	NS	NS	3.866	0.443	0.030	0.298
3.0–6.0	13	6	2.801	.0000	.0001	–0.577	.0000	.0005	4.133	0.843	–0.607	0.212
6.0–7.5	8	3	0.996	.0009	.0079	–0.284	.1436	NS	1.024	0.823	–0.204	0.482

TABLE 20. Horizontal Displacement of Infradentale per Half-Year, in mm (XID1)

		Males												Females			
Est. Age (Yrs.)	H. GRP	N	AGC	±CI	$\bar{v}$	$p(\bar{v}=0)$	$\bar{a}$	$p(\bar{a}=0)$	No. Ob.	$\bar{x}$	SD	$p(\bar{x}=0)$	No. Ob.	$\bar{x}$	SD	$p(\bar{x}=0)$
0.5	1								1	1.650			1	1.137		
1.0	2								9	0.767	0.588	.0045	4	0.707	1.128	NS
1.5	3								15	0.808	0.447	.0000	5	1.666	1.629	.0841
2.0	4	12	1.346	0.351	1.252	.0000			24	0.771	0.765	.0001	10	0.649	0.712	.0180
2.5	5	12	0.885	0.201	0.930	.0000	−0.322	NS	20	0.579	0.680	.0012	7	0.368	0.859	NS
3.0	6	12	0.424	0.353	0.235	NS	−0.695	.0384	22	0.575	1.018	.0150	13	0.088	0.889	NS
3.5	7	12	−0.037	0.613	0.110	NS	−0.125	NS	18	0.201	0.899	NS	10	0.126	0.694	NS
4.0	8	13	0.828	0.532	0.867	.0026			17	0.525	0.645	.0045	14	−0.022	0.613	NS
4.5	9	13	0.705	0.406	0.575	.0111	−0.292	NS	15	0.571	0.893	.0267	13	0.196	0.563	NS
5.0	10	13	0.582	0.301	0.684	.0295	0.109	NS	12	0.643	0.866	.0258	8	0.047	0.616	NS
5.5	11	13	0.459	0.247	0.532	.0033	−0.152	NS	13	0.358	0.624	.0609	6	−0.207	0.615	NS
6.0	12	13	0.337	0.276	0.242	.1138	−0.291	NS	12	0.291	0.592	.1167	12	0.061	0.661	NS
6.5	13	8	0.496	0.436	0.379	.0464			13	0.250	0.448	.0669	5	0.485	0.274	.0167
7.0	14	8	0.180	0.252	0.260	.0359	−0.414	.0152	7	0.283	0.370	.0899	1	−0.646		
7.5	15	8	−0.137	0.438	−0.155	NS	−0.120	NS	9	−0.150	0.371	NS	1	−0.041		
8.0	16								4	0.335	0.364	NS				
8.5	17								6	0.183	0.495	NS				
9.0	18								2	0.119						
9.5	19								3	0.292	0.509					
10.0	20								3	−0.329	0.564					

Summary Statistics (Males Only)

Age Interval (Years)	N	T	$\bar{v}$	$p(\bar{v}=0)$	$p(\mathbf{v}=0)$	$\bar{a}$	$p(\bar{a}=0)$	$p(\mathbf{a}=0)$	tau_0	±CI	tau_1	±CI
1.5–3.5	12	4	0.632	.0000	.0000	−0.381	.0121	.0053	1.346	0.265	−0.461	0.218
3.5–6.0	13	5	0.580	.0000	.0098	−0.156	.0183	NS	0.828	0.419	−0.123	0.118
6.0–7.5	8	3	0.161	.0364	.0443	−0.267	.0298	.0318	0.496	0.297	−0.317	0.243

TABLE 21. Vertical Displacement of Infradentale per Half-Year, in mm (YID1)

		Males											Females			
Est. Age (Yrs.)	H. GRP	N	AGC	±CI	$\bar{v}$	$p(\bar{v}=0)$	$\bar{a}$	$p(\bar{a}=0)$	No. Ob.	$\bar{x}$	SD	$p(\bar{x}=0)$	No. Ob.	$\bar{x}$	SD	$p(\bar{x}=0)$
0.5	1												1	0.168		
1.0	2								9	0.924	0.540	.0009	4	0.977	1.460	NS
1.5	3								15	0.100	0.545	NS	5	0.716	0.725	.0916
2.0	4	12	0.420	0.746	0.359	.1149			24	0.117	0.848	NS	10	0.230	0.696	NS
2.5	5	12	–0.206	0.390	–0.018	NS	–0.377	NS	20	–0.134	1.499	NS	7	0.689	0.809	.0652
3.0	6	12	–0.833	0.469	–1.396	.0014	–1.378	.0133	22	–0.646	1.466	.0512	13	–0.393	1.574	NS
3.5	7	12	–1.460	0.872	–0.959	.0624	0.437	NS	18	–0.438	1.094	.1078	10	0.084	1.539	NS
4.0	8	13	0.265	0.579	0.166	NS			17	–0.013	0.852	NS	14	–0.391	0.585	.0265
4.5	9	13	0.168	0.435	–0.088	NS	–0.254	NS	15	–0.037	0.915	NS	13	–0.127	0.356	NS
5.0	10	13	0.071	0.321	0.109	NS	0.197	NS	12	–0.049	0.688	NS	8	0.066	0.526	NS
5.5	11	13	–0.026	0.278	0.089	NS	–0.019	NS	13	0.078	0.734	NS	6	–0.025	0.398	NS
6.0	12	13	–0.123	0.335	–0.136	NS	–0.225	NS	12	–0.087	0.435	NS	12	–0.156	0.614	NS
6.5	13	8	0.176	0.429	0.296	NS			13	0.084	0.664	NS	5	0.394	0.909	NS
7.0	14	8	–0.044	0.158	–0.100	NS	–0.163	NS	7	–0.177	0.604	NS	1	–0.111		
7.5	15	8	–0.264	0.335	–0.263	.0158	–0.395	NS	9	–0.181	0.374	NS	1	–0.218		
8.0	16								4	0.572	0.596	.1510				
8.5	17								6	–0.022	0.663	NS				
9.0	18								2	0.133						
9.5	19								3	–0.035	0.490					
10.0	20								3	–0.388	0.301					

Summary Statistics (Males Only)

Age Interval (Years)	N	T	$\bar{v}$	$p(\bar{v}=0)$	$p(\mathbf{v}=0)$	$\bar{a}$	$p(\bar{a}=0)$	$p(\mathbf{a}=0)$	tau_0	±CI	tau_1	±CI
1.5–3.5	12	4	–0.503	.0002	.0042	–0.439	.0305	.0072	0.420	0.564	–0.627	0.367
3.5–6.0	13	5	0.028	NS	NS	–0.075	NS	NS	0.265	0.455	–0.097	0.137
6.0–7.5	8	3	–0.022	NS	NS	–0.279	.0877	.1330	0.176	0.293	–0.220	0.239

TABLE 22. Horizontal Displacement of Menton per Half-Year, in mm (XMN1)

	Males												Females			
Est. Age (Yrs.)	H. GRP	N	AGC	±CI	$\bar{v}$	$p(\bar{v}=0)$	$\bar{a}$	$p(\bar{a}=0)$	No. Ob.	$\bar{x}$	SD	$p(\bar{x}=0)$	No. Ob.	$\bar{x}$	SD	$p(\bar{x}=0)$
0.5	1								1	0.950			1	–0.402		
1.0	2								9	–0.216	0.474	NS	4	0.201	1.313	NS
1.5	3								15	–0.250	0.359	.0175	5	–0.353	0.473	NS
2.0	4	12	–0.557	0.570	–0.502	.0209			24	–0.278	0.639	.0442	10	–0.331	0.434	.0391
2.5	5	12	–0.728	0.464	–0.651	.0263	–0.149	NS	20	–0.653	0.757	.0011	7	0.072	0.576	NS
3.0	6	12	–0.898	0.523	–1.251	.0001	–0.600	.1394	22	–0.782	0.757	.0001	13	–0.693	0.749	.0059
3.5	7	12	–1.068	0.707	–0.928	.0004	0.323	NS	18	–0.745	0.824	.0013	10	–0.528	0.564	.0159
4.0	8	13	–1.007	0.429	–0.900	.0006			17	–0.800	0.769	.0006	14	–0.224	0.415	.0647
4.5	9	13	–0.818	0.303	–0.830	.0032	0.071	NS	15	–0.555	0.740	.0116	13	–0.122	0.362	NS
5.0	10	13	–0.630	0.234	–0.570	.0153	0.259	NS	12	–0.417	0.794	.0964	8	–0.209	0.416	NS
5.5	11	13	–0.441	0.271	–0.258	NS	0.312	NS	13	–0.388	0.843	.1225	6	–0.005	0.471	NS
6.0	12	13	–0.252	0.384	–0.247	.0842	0.011	NS	12	–0.132	0.582	NS	12	0.022	0.378	NS
6.5	13	8	0.089	0.371	0.053	NS			13	–0.162	0.434	NS	5	–0.038	0.533	NS
7.0	14	8	0.057	0.277	0.166	NS	–0.148	NS	7	0.047	0.509	NS	1	–0.549		
7.5	15	8	0.025	0.560	0.019	NS	0.114	NS	9	–0.007	0.744	NS	1	0.006		
8.0	16								4	0.070	0.238	NS				
8.5	17								6	0.234	0.694	NS				
9.0	18								2	–0.023						
9.5	19								3	–0.295	0.483					
10.0	20								3	–0.212	0.134					

Summary Statistics (Males Only)

Age Interval (Years)	N	T	$\bar{v}$	$p(\bar{v}=0)$	$p(\mathbf{v}=0)$	$\bar{a}$	$p(\bar{a}=0)$	$p(\mathbf{a}=0)$	tau_0	±CI	tau_1	±CI
1.5–3.5	12	4	–0.833	.0000	.0026	–0.142	.0779	.0674	–0.557	0.431	–0.170	0.219
3.5–6.0	13	5	–0.561	.0001	.0011	0.163	.0133	.0688	–1.007	0.337	0.189	0.131
6.0–7.5	8	3	0.079	NS	NS	–0.017	NS	NS	0.089	0.253	–0.032	0.263

TABLE 23. Vertical Displacement of Menton per Half-Year, in mm (YMN1)

	Males												Females			
Est. Age (Yrs.)	H. GRP	N	AGC	±CI	$\bar{v}$	$p(\bar{v}=0)$	$\bar{a}$	$p(\bar{a}=0)$	No. Ob.	$\bar{x}$	SD	$p(\bar{x}=0)$	No. Ob.	$\bar{x}$	SD	$p(\bar{x}=0)$
0.5	1								1	−0.643			1	0.299		
1.0	2								9	0.278	0.236	.0078	4	0.262	0.734	NS
1.5	3								15	0.436	0.459	.0025	5	0.348	0.153	.0070
2.0	4	12	0.519	0.144	0.528	.0001			24	0.358	0.281	.0000	10	0.345	0.308	.0062
2.5	5	12	0.528	0.121	0.527	.0000	0.000	NS	20	0.434	0.319	.0000	7	0.328	0.197	.0045
3.0	6	12	0.537	0.161	0.618	.0008	0.090	NS	22	0.578	0.450	.0000	13	0.252	0.303	.0111
3.5	7	12	0.545	0.234	0.545	.0000	−0.073	NS	18	0.352	0.373	.0009	10	0.096	0.151	.0735
4.0	8	13	0.787	0.449	0.795	.0000			17	0.618	0.449	.0000	14	0.098	0.356	NS
4.5	9	13	0.360	0.218	0.186	NS	−0.609	.0019	15	0.066	0.508	NS	13	0.099	0.209	.1127
5.0	10	13	0.129	0.169	0.090	NS	−0.095	NS	12	0.192	0.373	.1026	8	0.024	0.357	NS
5.5	11	13	0.092	0.165	0.113	.0159	0.022	NS	13	0.072	0.178	NS	6	−0.011	0.519	NS
6.0	12	13	0.252	0.216	0.203	.0125	0.090	NS	12	0.252	0.312	.0173	12	0.145	0.245	.0653
6.5	13	8	0.000	0.407	−0.026	NS			13	−0.041	0.257	NS	5	−0.323	0.731	NS
7.0	14	8	−0.017	0.185	0.143	NS	−0.193	NS	7	0.130	0.373	NS	1	0.023		
7.5	15	8	−0.034	0.293	−0.050	NS	0.169	NS	9	−0.018	0.218	NS	1	0.012		
8.0	16								4	−0.035	0.444	NS				
8.5	17								6	0.129	0.272	NS				
9.0	18								2	0.325						
9.5	19								3	−0.042	0.129					
10.0	20								3	−0.132	0.441					

Summary Statistics (Males Only)

Age Interval (Years)	N	T	$\bar{v}$	$p(\bar{v}=0)$	$p(\mathbf{v}=0)$	$\bar{a}$	$p(\bar{a}=0)$	$p(\mathbf{a}=0)$	tau_0	±CI	tau_1	±CI	tau_2	±CI
1.5–3.5	12	4	0.555	.0000	.0000	0.006	NS	NS	0.519	0.108	0.009	0.070		
3.5–6.0	13	5	0.277	.0000	.0002	−0.148	.0006	.0045	0.787	0.266	−0.524	0.220	0.098	0.042
6.0–7.5	8	3	0.021	NS	NS	−0.012	NS	NS	0.000	0.277	−0.017	0.205		

TABLE 24. Horizontal Displacement of Gonion per Half-Year, in mm (XGO1)

	Males												Females			
Est. Age (Yrs.)	H. GRP	N	AGC	±CI	$\bar{v}$	$p(\bar{v}=0)$	$\bar{a}$	$p(\bar{a}=0)$	No. Ob.	$\bar{x}$	SD	$p(\bar{x}=0)$	No. Ob.	$\bar{x}$	SD	$p(\bar{x}=0)$
0.5	1								1	3.605			1	2.644		
1.0	2								9	3.272	1.714	.0004	4	2.371	0.945	.0152
1.5	3								15	2.164	1.566	.0001	5	5.758	1.726	.0017
2.0	4	12	3.952	0.800	4.021	.0000			24	4.237	1.422	.0000	10	2.343	1.178	.0001
2.5	5	12	3.825	0.654	3.764	.0000	–0.258	NS	20	3.825	1.565	.0000	7	3.049	0.653	.0000
3.0	6	12	3.698	0.827	3.741	.0000	–0.023	NS	22	4.030	1.742	.0000	13	2.400	0.951	.0000
3.5	7	12	3.571	1.187	3.548	.0000	–0.193	NS	18	3.976	1.619	.0000	10	2.037	0.952	.0001
4.0	8	13	3.448	0.838	3.787	.0000			17	3.462	1.681	.0000	14	1.749	1.818	.0032
4.5	9	13	2.758	0.656	2.498	.0002	–1.290	.0305	15	2.213	1.273	.0000	13	1.545	0.899	.0000
5.0	10	13	2.068	0.515	1.876	.0000	–0.623	NS	12	1.886	1.044	.0001	8	1.445	2.189	.1041
5.5	11	13	1.378	0.455	1.156	.0003	–0.719	.0233	13	0.907	0.858	.0025	6	1.614	0.537	.0007
6.0	12	13	0.689	0.506	0.865	.0014	–0.292	NS	12	0.784	0.788	.0055	12	0.387	0.619	.0532
6.5	13	8	0.932	1.262	0.949	.0231			13	1.117	0.790	.0003	5	0.059	2.726	NS
7.0	14	8	0.702	0.856	0.663	.0481	–0.213	NS	7	0.699	0.940	.0967	1	–0.061		
7.5	15	8	0.473	0.843	0.451	.0969	–0.286	NS	9	0.414	1.400	NS	1	0.483		
8.0	16								4	0.158	0.319	NS				
8.5	17								6	–0.080	0.653	NS				
9.0	18								2	–0.155						
9.5	19								3	–0.020	0.129					
10.0	20								3	0.043	0.165					

Summary Statistics (Males Only)

Age Interval (Years)	N	T	$\bar{v}$	$p(\bar{v}=0)$	$p(\mathbf{v}=0)$	$\bar{a}$	$p(\bar{a}=0)$	$p(\mathbf{a}=0)$	tau_0	±CI	tau_1	±CI
1.5–3.5	12	4	3.769	.0000	.0000	–0.158	NS	NS	3.952	0.604	–0.127	0.365
3.5–6.0	13	5	2.036	.0000	.0000	–0.731	.0000	.0001	3.448	0.659	–0.690	0.182
6.0–7.5	8	3	0.688	.0184	NS	–0.249	NS	NS	0.932	0.861	–0.229	0.442

TABLE 25. Vertical Displacement of Gonion per Half-Year, in mm (YGO1)

	Males												Females			
Est. Age (Yrs.)	H. GRP	N	AGC	±CI	$\bar{v}$	$p(\bar{v}=0)$	$\bar{a}$	$p(\bar{a}=0)$	No. Ob.	$\bar{x}$	SD	$p(\bar{x}=0)$	No. Ob.	$\bar{x}$	SD	$p(\bar{x}=0)$
0.5	1								1	0.755			1	−0.330		
1.0	2								9	0.038	1.426	NS	4	0.381	1.375	NS
1.5	3								15	1.045	1.895	.0508	5	1.694	0.876	.0124
2.0	4	12	1.100	1.051	1.155	.0028			24	0.948	1.141	.0005	10	1.725	2.101	.0289
2.5	5	12	1.494	0.799	1.332	.0004	0.178	NS	20	1.388	1.338	.0002	7	1.606	1.536	.0326
3.0	6	12	1.888	0.815	2.043	.0001	0.711	.0801	22	1.881	1.264	.0000	13	0.608	0.661	.0062
3.5	7	12	2.281	1.086	2.237	.0003	0.194	NS	18	1.660	1.772	.0010	10	0.908	1.366	.0019
4.0	8	13	1.991	1.117	1.716	.0003			17	1.850	1.092	.0000	14	0.916	0.999	.0045
4.5	9	13	1.668	0.979	2.173	.0002	0.458	NS	15	2.303	1.490	.0000	13	0.366	1.510	NS
5.0	10	13	1.346	0.905	1.274	.0045	−0.899	.0931	12	1.181	1.380	.0129	8	0.829	1.622	NS
5.5	11	13	1.023	0.913	0.741	.0321	−0.532	NS	13	0.664	1.452	.1253	6	0.891	1.398	NS
6.0	12	13	0.707	1.000	0.727	.0353	−0.014	NS	12	0.667	1.394	.1254	12	−0.152	0.951	NS
6.5	13	8	0.420	1.462	0.305	NS			13	0.752	0.916	.0119	5	1.096	0.940	.0596
7.0	14	8	0.407	0.346	0.759	.1340	−0.443	NS	7	1.218	1.253	.0423	1	−0.223		
7.5	15	8	0.395	1.046	0.317	NS	0.455	NS	9	−0.172	1.908	NS	1	−1.002		
8.0	16								4	0.536	0.441	.0933				
8.5	17								6	−0.500	1.043	NS				
9.0	18								2	−0.194						
9.5	19								3	−0.177	0.722					
10.0	20								3	−0.004	0.225					

Summary Statistics (Males Only)

Age Interval (Years)	N	T	$\bar{v}$	$p(\bar{v}=0)$	$p(\mathbf{v}=0)$	$\bar{a}$	$p(\bar{a}=0)$	$p(\mathbf{a}=0)$	tau_0	±CI	tau_1	±CI
1.5–3.5	12	4	1.691	.0000	.0016	0.361	.0508	.1207	1.100	0.794	0.394	0.374
3.5–6.0	13	5	1.326	.0001	.0078	−0.247	.0163	.0239	1.991	0.879	−0.323	0.217
6.0–7.5	8	3	0.460	.0035	.0180	0.006	NS	NS	0.420	0.997	−0.013	0.834

Table 26. Vertical Displacement of Mandibular Alveolar Ridge per Half-Year, in mm (YMNALVR1)

		Males											Females			
Est. Age (Yrs.)	H. GRP	N	AGC	±CI	$\bar{v}$	$p(\bar{v}=0)$	$\bar{a}$	$p(\bar{a}=0)$	No. Ob.	$\bar{x}$	SD	$p(\bar{x}=0)$	No. Ob.	$\bar{x}$	SD	$p(\bar{x}=0)$
0.5	1												1	0.276		
1.0	2								9	0.339	1.627	NS	4	2.253	0.843	.0128
1.5	3								15	0.725	0.815	.0040	5	1.962	1.234	.0237
2.0	4	12	0.958	0.528	0.737	.0090			24	1.042	1.226	.0004	10	0.928	1.555	.0917
2.5	5	12	0.707	0.269	0.965	.0004	0.228	NS	20	0.757	1.261	.0147	7	0.456	1.052	NS
3.0	6	12	0.457	0.349	0.372	.0161	−0.593	.0128	22	0.408	0.656	.0083	13	0.261	0.772	NS
3.5	7	12	0.207	0.654	0.349	NS	−0.023	NS	18	0.429	1.200	.1464	10	0.509	0.733	.0557
4.0	8	13	0.739	0.486	0.861	.0002			17	0.875	1.558	.0342	14	0.384	0.794	.0937
4.5	9	13	0.528	0.305	0.489	.0753	−0.372	NS	15	0.300	1.161	NS	13	0.348	0.783	.1353
5.0	10	13	0.317	0.177	0.215	NS	−0.274	NS	12	0.454	1.162	NS	8	−0.026	1.313	NS
5.5	11	13	0.106	0.224	−0.219	NS	−0.434	.1549	13	−0.249	1.649	NS	6	0.564	0.539	.0504
6.0	12	13	−0.106	0.387	0.289	NS	0.508	NS	12	0.153	0.734	NS	12	0.278	0.814	NS
6.5	13	8	0.577	0.753	0.583	.0789			13	0.591	1.196	.1002	5	0.384	0.223	.0183
7.0	14	8	0.310	0.293	0.303	NS	−0.260	NS	7	0.593	0.856	.1381	1	0.041		
7.5	15	8	0.042	0.681	0.043	NS	−0.280	NS	9	0.153	0.949	NS	1	−0.128		
8.0	16								4	0.333	0.720	NS				
8.5	17								6	0.085	0.745	NS				
9.0	18								2	0.374						
9.5	19								3	0.160	0.359					
10.0	20								3	0.363	1.311					

Summary Statistics (Males Only)

Age Interval (Years)	N	T	$\bar{v}$	$p(\bar{v}=0)$	$p(\mathbf{v}=0)$	$\bar{a}$	$p(\bar{a}=0)$	$p(\mathbf{a}=0)$	tau_0	±CI	tau_1	±CI
1.5–3.5	12	4	0.605	.0000	.0003	−0.129	NS	.0808	0.958	0.399	−0.250	0.270
3.5–6.0	13	5	0.327	.0001	.0069	−0.143	.0421	.0443	0.739	0.382	−0.211	0.158
6.0–7.5	8	3	0.310	.0042	.0712	−0.270	NS	NS	0.577	0.514	−0.267	0.447

TABLE 27. Mandibular Length, in mm (MNLN)

	Males								Females			
Est. Age (Yrs.)	H. GRP	N	AGC	± CI	No. Ob.	$\bar{x}$	SD	± CI	No. Ob.	$\bar{x}$	SD	± CI
0.5	1				6	38.705	7.040	7.386	4	36.818	4.706	7.488
1.0	2				13	45.748	4.392	2.682	5	45.704	2.180	2.707
1.5	3	12	51.334	2.693	27	51.927	3.815	1.650	8	46.762	7.826	6.543
2.0	4	12	56.154	2.568	26	57.177	3.695	1.493	11	54.221	4.975	3.343
2.5	5	12	60.974	2.590	21	60.690	3.054	1.390	8	60.983	2.065	1.727
3.0	6	12	65.794	2.753	23	65.614	3.844	1.663	13	63.492	2.819	1.703
3.5	7	12	70.613	3.035	18	70.995	3.683	1.832	13	64.857	2.939	1.568
4.0	8	13	75.537	4.770	17	75.982	3.826	1.968	16	68.056	2.722	1.256
4.5	9	13	79.197	4.945	15	79.665	3.703	2.051	13	70.252	2.078	1.406
5.0	10	13	82.091	5.248	12	83.013	3.458	2.197	9	72.069	2.253	1.732
5.5	11	13	84.220	5.596	13	83.912	4.249	2.568	9	73.068	2.565	1.972
6.0	12	13	85.583	6.003	12	85.677	4.615	2.072	12	74.160	2.301	1.462
6.5	13	8	89.702	8.440	13	87.330	4.073	1.962	5	74.408	3.074	3.817
7.0	14	8	90.415	8.508	7	89.928	4.262	3.942				
7.5	15	8	91.128	8.768	9	89.913	4.045	3.109				
8.0	16				4	93.543	3.565	5.673				
8.5	17				6	93.736	3.029	3.179				
9.0	18				2	94.835						
9.5	19				3	93.740	3.958	9.832				
10.0	20				3	93.404	4.483	11.355				

Summary Statistics (Males Only)

Age Interval (Years)	N	T	tau_0	± CI	tau_1	± CI	tau_2	± CI
1.5–3.5	12	5	51.334	2.165	4.820	0.496		
3.5–6.0	13	6	71.111	2.945	4.809	0.962	–0.383	0.118
6.0–7.5	8	4	88.989	6.179	0.713	0.932		

TABLE 28. Change in Mandibular Length per Half-Year, in mm (MNLN1)

Est. Age (Yrs.)	H. GRP	Males N	AGC	±CI	$\bar{v}$	$p(\bar{v}=0)$	$\bar{a}$	$p(\bar{a}=0)$	No. Ob.	$\bar{x}$	SD	$p(\bar{x}=0)$	Females No. Ob.	$\bar{x}$	SD	$p(\bar{x}=0)$
0.5	1								2	11.285			2	10.261		
1.0	2								9	7.051	1.793	.0000	4	7.815	2.524	.0085
1.5	3								15	5.998	1.586	.0000	6	8.283	2.217	.0003
2.0	4	14	4.982	0.567	5.129	.0000			24	5.211	1.731	.0000	10	5.549	2.085	.0000
2.5	5	14	4.731	0.503	4.667	.0000	−0.461	NS	20	4.701	1.786	.0000	7	3.582	1.702	.0014
3.0	6	14	4.480	0.811	4.385	.0000	−0.283	NS	22	4.935	1.709	.0000	13	3.413	1.432	.0000
3.5	7	14	4.229	1.238	4.401	.0000	0.017	NS	18	5.124	1.349	.0000	10	2.898	1.401	.0001
4.0	8	13	4.426	1.107	4.948	.0000			17	4.446	1.069	.0000	15	2.004	1.158	.0000
4.5	9	13	3.660	0.909	3.971	.0000	−0.976	.0646	15	3.443	1.252	.0000	13	1.942	1.018	.0000
5.0	10	13	2.894	0.777	2.457	.0000	−1.514	.0043	12	2.491	0.944	.0000	8	1.192	0.431	.0001
5.5	11	13	2.129	0.749	2.108	.0000	−0.349	.1217	13	1.864	0.979	.0000	6	0.825	0.886	.0714
6.0	12	13	1.363	0.834	1.528	.0000	−0.581	.0436	12	1.687	0.977	.0001	12	1.293	1.133	.0023
6.5	13	8	1.463	1.527	1.487	.0035			13	1.340	1.033	.0005	13	1.230	0.816	.0280
7.0	14	8	1.017	1.022	1.386	.0062	−0.868	.0448	7	1.462	1.151	.0152				
7.5	15	8	0.571	0.920	0.519	.0436	−0.101	NS	9	0.645	0.664	.0195				
8.0	16								4	0.543	0.252	.0231				
8.5	17								6	0.438	0.566	.1162				
9.0	18								2	0.118						
9.5	19								3	0.198	0.484					
10.0	20								3	−0.071	0.690					

Summary Statistics (Males Only)

Age Interval (Years)	N	T	$\bar{v}$	$p(\bar{v}=0)$	$p(\mathbf{v}=0)$	$\bar{a}$	$p(\bar{a}=0)$	$p(\mathbf{a}=0)$	tau_0	±CI	tau_1	±CI
1.5–3.5	14	4	4.646	.0000	.0000	−0.242	NS	NS	4.982	0.430	−0.251	0.368
3.5–6.0	13	5	3.002	.0000	.0000	−0.855	.0000	.0001	4.426	0.871	−0.766	0.235
6.0–7.5	8	3	1.131	.0025	.0546	−0.484	.0240	.0839	1.463	1.042	−0.446	0.503

TABLE 29. Ramus Height, in mm (RMHT)

Est. Age (Yrs.)	H. GRP	N	AGC	±CI	No. Ob.	x̄	SD	±CI	No. Ob.	x̄	SD	±CI
	Males								Females			
0.5	1				6	16.309	5.436	5.705	4	15.572	3.406	5.420
1.0	2				13	21.684	2.875	1.738	5	22.547	1.774	2.203
1.5	3	12	26.696	0.965	23	25.987	1.785	0.772	8	22.640	5.267	4.404
2.0	4	12	29.064	1.096	26	28.361	1.899	0.767	11	27.390	3.452	2.319
2.5	5	12	31.433	1.420	21	30.848	2.289	1.043	8	30.559	1.273	1.064
3.0	6	12	33.802	1.837	23	32.986	2.508	1.085	13	32.387	2.427	1.467
3.5	7	12	36.170	2.298	18	35.586	2.559	1.273	13	33.228	2.220	1.342
4.0	8	13	39.112	4.367	17	38.416	2.399	1.234	16	34.685	2.421	1.291
4.5	9	13	40.809	4.862	15	40.174	2.677	1.482	13	36.547	2.327	1.406
5.0	10	13	42.094	5.346	12	40.924	3.336	2.120	9	36.983	2.428	1.866
5.5	11	13	42.967	5.665	13	41.855	3.327	2.011	9	38.129	3.139	2.413
6.0	12	13	43.429	5.793	12	42.785	3.454	2.195	12	39.019	2.806	1.783
6.5	13	8	42.142	6.077	13	42.928	3.325	2.009	5	38.344	4.152	5.156
7.0	14	8	42.766	6.468	7	43.845	4.531	4.191				
7.5	15	8	43.390	6.882	9	42.812	4.612	3.545				
8.0	16				4	44.376	2.212	3.521				
8.5	17				6	44.124	1.797	1.886				
9.0	18				2	46.479						
9.5	19				3	45.218	2.847	7.072				
10.0	20				3	43.515	2.882	7.160				

Summary Statistics (Males Only)

Age Interval (Years)	N	T	tau_0	±CI	tau_1	±CI	tau_2	±CI
1.5–3.5	12	5	26.296	0.776	2.369	0.420		
3.5–6.0	13	6	37.003	2.532	2.314	1.154	–0.206	0.130
6.0–7.5	8	4	41.518	4.121	0.624	0.402		

TABLE 30. Change in Ramus Height per Half-Year, in mm (RMHT1)

		Males											Females			
Est. Age (Yrs.)	H. GRP	N	AGC	±CI	$\bar{v}$	$p(\bar{v}=0)$	$\bar{a}$	$p(\bar{a}=0)$	No. Ob.	$\bar{x}$	SD	$p(\bar{x}=0)$	No. Ob.	$\bar{x}$	SD	$p(\bar{x}=0)$
0.5	1												2	7.936		
1.0	2								9	5.699	1.253	.0000	4	4.033	2.989	.0739
1.5	3								15	3.603	2.182	.0000	6	6.676	4.102	.0105
2.0	4	12	2.621	0.759	2.789	.0000			24	2.381	1.751	.0000	10	3.686	1.768	.0001
2.5	5	12	2.494	0.550	2.269	.0001	−0.520	NS	20	2.709	1.337	.0000	7	1.105	1.729	.1418
3.0	6	12	2.366	0.462	2.327	.0000	0.057	NS	22	2.199	1.600	.0000	13	2.255	1.372	.0001
3.5	7	12	2.239	0.557	2.294	.0000	−0.033	NS	18	2.431	1.426	.0000	10	0.982	1.265	.0365
4.0	8	13	2.108	1.317	2.043	.0000			17	1.974	1.379	.0000	15	1.407	1.315	.0010
4.5	9	13	1.697	1.031	1.957	.0000	−0.086	NS	15	1.884	0.856	.0000	13	1.475	1.116	.0005
5.0	10	13	1.285	0.760	0.532	.0903	−1.425	.0004	12	0.471	1.293	NS	8	−0.627	2.033	NS
5.5	11	13	0.873	0.610	1.320	.0003	0.788	.0613	13	1.163	1.347	.0089	6	0.771	1.438	NS
6.0	12	13	0.462	0.595	0.499	.0160	−0.822	.0396	12	0.579	1.168	.1139	12	0.259	1.193	NS
6.5	13	8	0.980	0.764	0.962	.0130			13	0.690	0.766	.0070	5	0.609	2.097	NS
7.0	14	8	0.599	0.403	0.621	.0690	1.033	NS	7	0.356	1.217	NS				
7.5	15	8	0.218	0.844	0.202	NS	2.167	NS	9	0.550	1.713	NS	1	0.563		
8.0	16								4	−0.023	0.609	NS				
8.5	17								6	0.215	0.382	NS				
9.0	18								2	0.323						
9.5	19								3	0.101	0.385					
10.0	20								3	−0.498	0.605					

Summary Statistics (Males Only)

Age Interval (Years)	N	T	$\bar{v}$	$p(\bar{v}=0)$	$p(\mathbf{v}=0)$	$\bar{a}$	$p(\bar{a}=0)$	$p(\mathbf{a}=0)$	tau_0	±CI	tau_1	±CI
1.5–3.5	12	4	2.420	.0000	.0000	−0.165	.1278	NS	2.621	0.573	−0.127	0.230
3.5–6.0	13	5	1.270	.0000	.0016	−0.386	.0006	.0019	2.108	1.036	−0.412	0.260
6.0–7.5	8	3	0.595	.0011	.0143	−0.380	.0680	NS	0.980	0.521	−0.381	0.475

TABLE 31. Change in Angle between Occlusal Plane and Maxilla per Half-Year, in Degrees (MXRLOCPL1)

	Males												Females			
Est. Age (Yrs.)	H. GRP	N	AGC	±CI	$\bar{v}$	$p(\bar{v}=0)$	$\bar{a}$	$p(\bar{a}=0)$	No. Ob.	$\bar{x}$	SD	$p(\bar{x}=0)$	No. Ob.	$\bar{x}$	SD	$p(\bar{x}=0)$
0.5	1												1	1.550		
1.0	2								9	–1.871	6.550	NS	4	5.184	5.069	.1333
1.5	3								15	–1.114	7.332	NS	4	4.126	4.473	.1623
2.0	4	11	0.155	3.742	1.269	NS			24	2.546	5.233	.0330	10	3.035	4.157	.0436
2.5	5	11	1.097	1.961	0.015	NS	2.890	.0398	20	0.157	3.817	NS	6	–0.661	2.849	NS
3.0	6	11	2.039	3.163	2.905	.0083	–1.254	NS	22	2.970	3.312	.0004	8	0.619	3.339	NS
3.5	7	13	0.580	1.333	1.852	.0339			18	1.014	3.636	NS	10	–0.666	3.835	NS
4.0	8	13	0.495	1.000	1.112	.1529	–0.740	NS	17	0.648	2.767	NS	14	0.114	4.976	NS
4.5	9	13	0.411	0.711	0.524	NS	–0.588	NS	15	0.326	3.713	NS	13	0.864	4.129	NS
5.0	10	13	0.326	0.541	0.023	NS	–0.500	NS	12	0.245	1.966	NS	8	–0.061	3.858	NS
5.5	11	13	0.242	0.600	1.074	.0084	1.051	.1130	13	1.173	1.905	.0464	6	–0.610	4.365	NS
6.0	12	13	0.157	0.843	0.072	NS	–1.002	NS	12	0.211	1.706	NS	12	–0.082	1.888	NS
6.5	13	8	0.675	2.226	0.632	NS			13	0.109	1.521	NS	5	0.662	1.459	NS
7.0	14	8	–0.126	1.314	–0.506	NS	–0.110	NS	7	–0.456	1.781	NS	1	–1.405		
7.5	15	8	–0.926	1.149	–0.616	NS	–1.139	.0630	9	–0.854	2.863	NS	1	0.318		
8.0	16								4	0.467	0.740	NS				
8.5	17								6	–0.153	2.403	NS				
9.0	18								2	0.339						
9.5	19								3	0.834	1.851					
10.0	20								3	–1.550	0.221					

Summary Statistics (Males Only)

Age Interval (Years)	N	T	$\bar{v}$	$p(\bar{v}=0)$	$p(\mathbf{v}=0)$	$\bar{a}$	$p(\bar{a}=0)$	$p(\mathbf{a}=0)$	tau_0	±CI	tau_1	±CI
1.5–3.5	11	3	1.396	.0187	.0789	0.818	NS	.1358	0.155	2.566	0.942	1.959
3.0–6.0	13	6	0.776	.0046	.1258	–0.356	.0123	.1529	0.580	1.002	–0.085	0.310
6.0–7.5	8	3	–0.163	NS	.1073	–0.624	.1295	.0942	0.675	1.518	–0.080	0.810

TABLE 32. Change in Angle between Occlusal Plane and Mandible per Half-Year, in Degrees (MRLOCPL1)

	Males												Females			
Est. Age (Yrs.)	H. GRP	N	AGC	±CI	$\bar{v}$	$p(\bar{v}=0)$	$\bar{a}$	$p(\bar{a}=0)$	No. Ob.	$\bar{x}$	SD	$p(\bar{x}=0)$	No. Ob.	$\bar{x}$	SD	$p(\bar{x}=0)$
0.5	1								1	10.245			1	8.302		
1.0	2								9	1.090	1.866	.1180	4	1.327	1.708	NS
1.5	3								15	1.417	3.375	.1264	5	2.677	1.660	.0226
2.0	4	11	1.116	2.471	1.980	.0308			22	1.975	2.500	.0013	10	1.927	2.401	.0318
2.5	5	11	1.510	1.403	0.953	.0712	1.118	.1321	20	0.468	2.476	NS	6	0.541	1.527	NS
3.0	6	11	1.904	2.711	2.072	.0122	−1.027	NS	22	1.967	2.808	.0035	8	1.425	2.742	NS
3.5	7	13	1.347	2.487	2.364	.0125			18	1.083	2.783	.1170	10	−0.267	2.503	NS
4.0	8	13	1.072	1.845	1.532	.0888	−0.832	NS	17	1.311	2.833	.0745	14	0.487	1.778	NS
4.5	9	13	0.794	1.232	0.560	NS	−0.972	NS	15	0.218	2.277	NS	13	0.587	2.129	NS
5.0	10	13	0.517	0.722	−0.153	NS	−0.713	NS	12	−0.108	1.087	NS	8	−0.122	2.068	NS
5.5	11	13	0.240	0.646	0.994	.0257	1.146	.0736	13	0.825	1.471	.0661	6	−0.318	2.099	NS
6.0	12	13	−0.037	1.098	−0.064	NS	−1.057	NS	12	0.159	2.307	NS	12	−0.040	1.308	NS
6.5	13	8	0.850	1.803	1.228	.0854			13	0.746	1.590	.1166	5	1.149	1.752	NS
7.0	14	8	0.302	0.659	−0.565	NS	0.107	NS	7	−1.085	2.640	NS	1	−1.130		
7.5	15	8	−0.247	2.011	−0.458	NS	−1.792	NS	9	−0.165	1.070	NS	1	0.855		
8.0	16								4	−0.306	0.125	.0163				
8.5	17								6	−0.061	1.166	NS				
9.0	18								2	−0.007						
9.5	19								3	0.454	0.935					
10.0	20								3	0.142	0.318					

Summary Statistics (Males Only)

Age Interval (Years)	N	T	$\bar{v}$	$p(\bar{v}=0)$	$p(\mathbf{v}=0)$	$\bar{a}$	$p(\bar{a}=0)$	$p(\mathbf{a}=0)$	tau_0	±CI	tau_1	±CI
1.5–3.5	11	3	1.668	.0009	.0204	0.046	NS	NS	1.116	1.694	0.394	1.496
3.0–6.0	13	6	0.872	.0011	.0796	−0.486	.0119	.0852	1.349	2.052	−0.277	0.551
6.0–7.5	8	3	0.069	NS	NS	−0.843	NS	NS	0.850	1.230	−0.548	1.223

REFERENCES

Alt, F. B. 1982. Bonferroni inequalities and intervals. In *Encyclopedia of Statistical Sciences*, vol. 1, ed. S. Kotz and N. L. Johnson. New York: Wiley.

Baer, M. J., and S. K. Nanda. 1977. A Commentary of the Growth and Form of the Cranial Base. Reprinted from *Symposium on Development of the Basicranium*, ed. J. F. Bosma. DHEW Publication No. (NIH) 77-989.

Baume, L. J. 1951a. The postnatal growth of the maxilla in *Macaca mulatta*. *J. Dent. Res.* 30: 501–502 (abstract).

Baume, L. J. 1951b. The postnatal growth of the mandible in *Macaca mulatta*. *J. Dent. Res.* 30: 502 (abstract).

Baume, L. J. 1951c. The relationship between jaw growth, tooth development and tooth eruption. *J. Dent. Res.* 30: 502–503 (abstract).

Behrents, R. G. 1985. *Growth in the Aging Craniofacial Skeleton*. Monograph 17, Craniofacial Growth Series. Center for Human Growth and Development, The University of Michigan, Ann Arbor.

Björk, A. 1955a. Facial growth in man, studied with the aid of metallic implants. *Acta Odont. Scand.* 13: 9–34.

Björk, A. 1955b. Cranial base development: A follow-up x-ray study of the individual variation in growth occurring between the age of 12 and 20 years and its relation to brain case and face development. *Am. J. Orthod.* 41: 198–225.

Björk, A. 1963. Variations in the growth pattern of the human mandible: Longitudinal radiographic study by the implant method. *J. Dent. Res.* (suppl. to no. 1) 42: 400–411.

Björk, A. 1966. Sutural growth of the upper face studied by the implant method. *Acta Odont. Scand.* 24: 109–127.

Björk, A. 1968. The use of metallic implants in the study of facial growth in children: Method and application. *Am. J. Phys. Anthropol.* 29: 243–254.

Björk, A. 1969. Prediction of growth rotation. *Am. J. Orthod.* 55: 587–599.

Björk, A., and V. Skieller. 1972. Facial development and tooth eruption: An implant study at the age of puberty. *Am. J. Orthod.* 62: 339–383.

Björk, A., and V. Skieller. 1977. Growth of the maxilla in three dimensions as revealed radiographically by the implant method. *Brit. J. of Orthod.* 4: 53–64.

Boersma, H., F.P.G.M. van der Linden, and B. Prahl-Andersen. 1979. Craniofacial development. In *A Mixed Longitudinal Interdisciplinary*

Study of Growth and Development, ed. B. Prahl-Andersen, C. J. Kowalski, and P.H.J.M. Heydendael. New York: Academic Press.

Bookstein, F. L. 1983. Measuring treatment effects on craniofacial growth. In *Clinical Alteration of the Growing Face*, ed. J. A. McNamara, K. A. Ribbens, and R. P. Howe. Monograph 14, Craniofacial Growth Series. Center for Human Growth and Development, The University of Michigan, Ann Arbor.

Bosscher, G. P. 1985. Adaptations in the Maxillary Complex Induced by Alterations in Muscle Length. Master's thesis, Department of Orthodontics, The University of Michigan, Ann Arbor.

Bravo, L. A., I. L. Nielsen, and A. J. Miller. 1989. Changes in facial morphology in *Macaca mulatta*, a longitudinal cephalometric study from 1.5 to 5 years. *Am. J. Orthod. Dentofac. Orthop.* 96: 26–35.

Broadbent, B. H., Sr., B. H. Broadbent, Jr., and W. H. Golden. 1975. *Bolton Standards of Dentofacial Developmental Growth.* St. Louis, Mo.: C. V. Mosby Co.

Carlson, D. S. 1983. Growth of the masseter muscle in rhesus monkeys (*Macaca mulatta*). *Am. J. Phys. Anthropol.* 60: 401–410.

Carlson, D. S. 1985. *Introduction to Craniofacial Biology: Growth and Adaptation in the Craniofacial Complex.* Validation edition, no. 1. The University of Michigan School of Dentistry, Ann Arbor.

Carlson, D. S., and E. D. Schneiderman. 1983. Cephalometric analysis of adaptations after lengthening of masseter muscle in adult rhesus monkeys, *Macaca mulatta*. *Arch. Oral Biol.* 28: 627–637.

Carlson, D. S., E. Ellis, E. D. Schneiderman, and J. C. Ungerleider. 1982. Experimental models of surgical intervention in the growing face: Cephalometric analysis of facial growth and relapse. In *The Effect of Surgical Intervention on Craniofacial Growth*, ed. J. A. McNamara, Jr., D. S. Carlson, and K. A. Ribbens. Monograph 12, Craniofacial Growth Series. Center for Human Growth and Development, The University of Michigan, Ann Arbor.

Carlson, D. S., J. A. McNamara, and D. H. Jaul. 1978. Histological analysis of the growth of the mandibular condyle in the rhesus monkey (*Macaca mulatta*). *Am. J. Anat.* 151: 103–118.

Carlson, D. S., P. A. Nemeth, and P. C. Dechow. 1985. Fiber and sarcomere length in digastric muscle after mandibular advancement. *J. Dent. Res.* 64: 367 (abstract).

Cheverud, J. M. 1981. Epiphyseal union and dental eruption in *Macaca mulatta*. *Am. J. Phys. Anthropol.* 56: 157–167.

Cheverud, J. M., J. L. Lewis, W. Bachrach, and W. D. Lew. 1983. The measurement of form and variation in form: An application of three-dimensional quantitative morphology by finite-element methods. *Am. J. Phys. Anthropol.* 62: 151–165.

Comas, J. 1960. *Manual of Physical Anthropology.* Springfield, Ill.: Chas. C. Thomas.

Copray, J.C.V.M., H.W.B. Jansen, and H. S. Duterloo. 1983. Growth of the mandibular condylar cartilage of the rat in serum-free organ culture. *Arch. Oral Biol.* 28: 967–974.

Copray, J.C.V.M., H.W.B. Jansen, and H. S. Duterloo. 1985. The role of biomechanical factors in mandibular condylar cartilage growth and remodeling *in vitro*. In *Developmental Aspects of Temporomandibular Joint Dysfunction*, ed. D. S. Carlson, J. A. McNamara, and K. A. Ribbens. Monograph 16, Craniofacial Growth Series. Center for Human Growth and Development, The University of Michigan, Ann Arbor.

Craven, A. H. 1956. Growth in width of the head of the rhesus monkey as revealed by vital staining. *Am. J. Orthod.* 42: 341–362.

Dechow, P. C., and D. S. Carlson. 1982a. Bite force and gape in rhesus monkeys. *Am. J. Phys. Anthropol.* 57: 179 (abstract).

Dechow, P. C., and D. S. Carlson. 1982b. Development of masticatory muscle force in macaques. *J. Dent. Res.* 61: 211 (abstract).

Dechow, P. C., D. S. Carlson, J. P. LaBanc, and B. N. Epker. 1983. Changes in bite force by masticatory muscle repositioning in *Macaca nemestrina*. *Am. J. Phys. Anthropol.* 60: 187–188 (abstract).

Diaconis, P., and B. Efron. 1983. Computer intensive methods in statistics. *Sci. Amer.* 248: 116–130.

Duterloo, H. S., and D. H. Enlow. 1970. A comparative study of cranial growth in *Homo* and *Macaca*. *Am. J. Anat.* 127: 357–368.

Edlefsen, L. E., and S. D. Jones. 1985. *GAUSS*. Kent, Wash.: Applied Technical Systems.

Elgoyen, J. C., M. L. Riolo, L. W. Graber, R. E. Moyers, and J. A. McNamara. 1972. Craniofacial growth in juvenile *Macaca mulatta*: A cephalometric study. *Am. J. Phys. Anthropol.* 36: 369–376.

Elston, R. C., and J. E. Grizzle. 1962. Estimation of time-response curves and their confidence bands. *Biometrics*, 18: 148–159.

Enlow, D. H. 1966. A comparative study of facial growth in *Homo* and *Macaca*. *Am. J. Phys. Anthropol.* 24: 293–308.

Enlow, D. H. 1968. *The Human Face*. New York: Hoeber Medical Division, Harper & Rowe.

Enlow, D. H. 1982. *Handbook of Facial Growth*. 2d ed. Philadelphia: W. B. Saunders Company.

Enlow, D. H., T. Kuroda, and A. B. Lewis. 1971. Intrinsic craniofacial compensations. *Angle Orthod.* 41: 271–285.

Erickson, L. C. 1958. Facial growth in the macaque monkey: A longitudinal roentgenographic study using metallic implants. Master's thesis, University of Washington, Seattle.

Fox, D. J., and K. E. Guire. 1976. *Documentation for MIDAS*. Statistical Research Laboratory, The University of Michigan, Ann Arbor.

Gans, B. J., and B. G. Sarnat. 1951. Sutural facial growth of the Macaca rhesus monkey: A gross and serial roentgenographic study by means of metallic implants. *Am. J. Orthod.* 37: 827–841.

Gaven, J. A. 1953. Growth and development of the chimpanzee: A longitudinal and comparative study. *Human Biol.* 25: 93–143.

Gaven, J. A., and T. C. Hutchinson. 1973. The problem of age estimation: A study using rhesus monkeys *Macaca mulatta*. *Am. J. Phys. Anthropol.* 38: 69–82.

Healy, M.J.R. 1969. Rao's paradox concerning multivariate tests of significance. *Biometrics* 25: 411–413.

Hills, M. 1968. A note on the analysis of growth curves. *Biometrics* 24: 189–196.

Hoel, P. G. 1964. Methods for comparing growth type curves. *Biometrics* 20: 859–872.

Hurme, V. O., and G. van Wagenen. 1953. Basic data on the emergence of deciduous teeth in the monkey *Macaca mulatta*. *Proc. Amer. Philos. Soc.* 97: 291–315.

Hurme, V. O., and G. van Wagenen. 1956. Emergence of permanent first molars in the monkey *Macaca mulatta*: Association with other growth phenomena. *Yale J. Biol. and Med.* 28: 538–567.

Hurme, V. O., and G. van Wagenen. 1961. Basic data on the emergence of permanent teeth in the rhesus monkey *Macaca mulatta*. *Proc. Amer. Philos. Soc.* 105: 105–140.

Israelsohn, W. J. 1960. Description and modes of analysis of human growth. In *Human Growth*, vol. 3, ed. J. M. Tanner. Oxford: Pergamon.

Kanouse, M. C., S. P. Ramfjord, and C. E. Nasjleti. 1969. Condylar growth in rhesus monkeys. *J. Dent. Res.* 48: 1171–1176.

King, A. H. 1990. Craniofacial Morphology of the Cayo Santiago Rhesus Macaque (*Macaca mulatta*). Ph.D. diss., Department of Physical Anthropology, State University of New York at Buffalo, New York.

King, A. H., and E. D. Schneiderman. 1991. Comparison of craniofacial dimensions in free-ranging and laboratory rhesus macaques (*Macaca mulatta*). *Amer. J. Phys. Anthrop.* 84 (Suppl. 12): 105 (abstract).

Kowalski, C. J. 1972. A commentary on the use of multivariate statistical methods in anthropometric research. *Am. J. Phys. Anthropol.* 36: 119–132.

Kowalski, C. J., and K. E. Guire. 1974. Longitudinal data analysis. *Growth*, 38: 131–169.

Kraus, B. S., R. E. Jordan, and L. Abrams. 1969. *A Study of the Masticatory System: Dental Anatomy and Occlusion*. Baltimore, Md.: The Williams and Wilkins Company.

Krogman, W. M., and V. Sassouni. 1957. *Syllabus in roentgenographic cephalometry*. Philadelphia Center for Research in Child Growth, Philadelphia (reprinted 1973 by University Microfilms, Ann Arbor, Michigan).

Lestrel, P., and J. E. Sirianni. 1982. The cranial base in *Macaca nemestrina*: Shape changes during adolescence. *Human Biol.* 54: 7–21.

Luder, H. U. 1987a. Evidence for a pubertal growth spurt in mandibular condylar growth of nonhuman primates. In *Craniofacial Growth During Adolescence*, ed. D. S. Carlson and K. A. Ribbens. Monograph 20,

Craniofacial Growth Series. Center for Human Growth and Development, The University of Michigan, Ann Arbor.
Luder, H. U. 1987b. Structure and growth activities of the mandibular condyle in monkeys (*Macaca fascicularis*): II. Synergistic behavior of cell dynamics and metabolism. *Am. J. Anat.* 178: 185–192.
McNamara, J. A., Jr. 1972. *Neuromuscular and Skeletal Adaptations to Altered Orofacial Function.* Monograph 1, Craniofacial Growth Series. Center for Human Growth and Development, The University of Michigan, Ann Arbor.
McNamara, J. A., Jr., and L. W. Graber. 1975. Mandibular growth in the rhesus monkey *Macaca mulatta. Am. J. Phys. Anthropol.* 42: 15–24.
McNamara, J. A., Jr., M. L. Riolo, and D. H. Enlow. 1976. Growth of the maxillary complex in the rhesus monkey *Macaca mulatta. Am. J. Phys. Anthropol.* 44: 15–26.
Martin, R. 1928. *Lehrbuch der Anthropologie.* 2d ed. 3 vols. Jena: G. Fischer.
Michejda, M., and S. Weinstein. 1971. Adaptive growth changes of the gonial region in *Macaca mulatta. Am. J. Phys. Anthropol.* 34: 133–142.
MIDAS. 1976. *Elementary Statistics Using MIDAS.* Statistical Research Laboratory, The University of Michigan, Ann Arbor.
Moore, A. W. 1949. Head growth of the macaque monkey as revealed by vital staining, embedding, and undecalcified sectioning. *Am. J. Orthod.* 35: 654–671.
Moss, M. L., R. Skalak, H. Patel, K. Sen, L. Moss-Salentijn, M. Shinozuka, and H. Vilmann. 1985. Finite element method modeling of craniofacial growth. *Am. J. Orthod.* 87: 453–462.
Moyers, R. E., and F. L. Bookstein. 1979. The inappropriateness of conventional cephalometrics. *Am. J. Orthod.* 75: 599–617.
Nemeth, P. A., P. C. Dechow, and D. S. Carlson. 1983. Changes in sarcomere length and bite force with increasing gape in the masticatory muscles of *Macaca fasicularis. Am. J. Phys. Anthropol.* 60: 231 (abstract).
Neter, J., and W. Wasserman. 1974. *Applied Linear Statistical Models: Regression, Analysis of Variance, and Experimental Designs.* Homewood, Ill.: Richard D. Irwin, Inc.
Nielsen, I. L., L. A. Bravo, and A. J. Miller. 1989a. Normal maxillary and mandibular growth and development in the Macaca mulatta. *J. Dent. Res.* 68: (S.I.) 293 (abstract).
Nielsen, I. L., L. A. Bravo, and A. J. Miller. 1989b. Normal maxillary and mandibular growth and dentoalveolar development in *Macaca mulatta,* a longitudinal cephalometric study from 2 to 5 years. *Am. J. Orthod. Dentofac. Orthop.* 96: 405–415.
Olshan, A. F., A. F. Siegel, and D. R. Swindler. 1982. Robust and least-squares orthogonal mapping: Methods for the study of cephalofacial form and growth. *Am. J. Phys. Anthropol.* 59: 131–137.
Petrovic, A. G., J. J. Stutzmann, and N. Gasson. 1981. The final length of the mandible: is it genetically predetermined. In *Craniofacial Biology,*

ed. D. S. Carlson. Monograph 10, Craniofacial Growth Series. Center for Human Growth and Development, The University of Michigan, Ann Arbor.

Pihl, E. B. 1959. A serial study of the growth of various cranial and facial bones in the macaque monkey. Master's thesis, University of Washington, Seattle.

Potthoff, R. F., and S. N. Roy. 1964. A generalized multivariate analysis model useful especially for growth curve problems. *Biometrika* 51: 313–326.

Rao, C. R. 1959. Some problems involving linear hypotheses in multivariate analysis. *Biometrika*, 46: 49–58.

Rao, C. R. 1966. Covariance adjustment and related problems in multivariate analysis. In *Multivariate Analysis*, ed. P. R. Krishnaiah. New York: Academic Press.

Riolo, M. L., R. E. Moyers, J. A. McNamara, Jr., and W. S. Hunter. 1974. *An Atlas of Craniofacial Growth: Cephalometric Standards from the University School Growth Study, The University of Michigan.* Monograph 2, Craniofacial Growth Series. Center for Human Growth and Development, The University of Michigan, Ann Arbor.

Robinson, I. B., and B. G. Sarnat. 1955. Growth pattern of the pig mandible. A serial roentgenographic study using metallic implants. *Am. J. Anat.* 96: 37–64.

Roy, E. W., and B. G. Sarnat. 1956. Growth in the length of rabbit ribs at the costochondral junction. *Surg. Gynecol. and Obst.* 103: 481–486.

Sarnat, B. G. 1968. Growth of bones as revealed by implant markers in animals. *Am. J. Phys. Anthropol.* 29: 255–286.

SAS Institute, Inc. 1982a. *SAS User's Guide: Basics. 1982 Edition.* Cary, N.C.: SAS Institute.

SAS Institute, Inc. 1982b. *SAS User's Guide: Statistics. 1982 Edition.* Cary, N.C.: SAS Institute.

Schneiderman, E. D. 1985. A Longitudinal Cephalometric Study of Normal Craniofacial Growth of the Rhesus Monkey (*Macaca mulatta*). Ph.D. diss., Department of Anthropology, The University of Michigan, Ann Arbor.

Schneiderman, E. D., and D. S. Carlson. 1981. Growth and remodeling of the mandible following alteration of function in *adult* rhesus monkeys. *Am. J. Phys. Anthropol.* 54: 275 (abstract).

Schneiderman, E. D., and D. S. Carlson. 1983. Condylar adaptations to altered mandibular position in adult rhesus monkeys. *J. Dent. Res.* 62: 284 (abstract).

Schneiderman, E. D., and D. S. Carlson. 1985. Cephalometric analysis of condylar adaptations to altered mandibular position in adult rhesus monkeys (*Macaca mulatta*). *Arch. Oral Biol.* 30: 49–54.

Schneiderman, E. D., and C. J. Kowalski. 1985. Implementation of Rao's polynomial growth curve model using SAS. *Am. J. Phys. Anthropol.* 67: 323–333.

Schneiderman, E. D., and C. J. Kowalski. 1989. Implementation of Hills' growth curve analysis for unequal-time intervals using *GAUSS*. *Am. J. Hum. Biol.* 1: 31–42.

Schneiderman, E. D., and C. J. Kowalski. In prep. Implementation of the Rao/Hills polynomial growth curve model extended to G-groups.

Schneiderman, E. D., A. H. King, and C. J. Kowalski. 1991. Error involved in using cross-sectional statistics on longitudinal craniofacial data. *J. Dent. Res.* 70(SI): 334 (abstract).

Schneiderman, E. D., C. J. Kowalski, and T. R. Ten Have. 1990. A *GAUSS* program for computing an index of tracking from longitudinal observations. *Am. J. Hum. Biol.* 2: 475–490.

Schneiderman, E. D., S. M. Willis, and C. J. Kowalski. In prep. Implementation of randomization tests for polynomial growth curves using *GAUSS*. I. The completely randomized design.

Selman, A. J., and B. G. Sarnat. 1953. A head-holder for serial roentgenography of the rabbit skull. *Anat. Rec.* 115: 627–634.

Selman, A. J., and B. G. Sarnat. 1957. Growth of the rabbit snout after the extirpation of the frontonasal suture: A gross and serial roentogenographic study by means of metallic implants. *Am. J. Anat.* 101: 273–294.

Shimshoni, Z., I. Binderman, N. Fine, and D. Somjen. 1984. Mechanical and hormonal stimulation of cell cultures derived from young rat mandible condyle. *Arch. Oral Biol.* 29: 827–831.

Siegel, A. F., and R. H. Benson. 1982. A robust comparison of biological shapes. *Biometrics* 38: 341.

Sirianni, J. E. 1985. Nonhuman primates as models for human craniofacial growth. In *Nonhuman Primate Models for Human Growth and Development*, ed. E. Watts. New York: Alan R. Liss, Inc.

Sirianni, J. E., and D. R. Swindler. 1979. A review of postnatal craniofacial growth in Old World monkeys and apes. In *Yearbook of Physical Anthropology*, vol. 22, ed. K. A. Bennett. Washington, D.C.: American Association of Physical Anthropologists.

Sirianni, J. E., and D. R. Swindler. 1985. *Growth and Development of the Pigtailed Macaque*. Boca Raton, Fla.: CRC Press.

Sirianni, J. E., and A. L. Van Ness. 1978. Postnatal growth of the cranial base in *Macaca nemestrina*. *Am. J. Phys. Anthropol.* 49: 329–340.

Sirianni, J. E., A. L. Van Ness, and D. R. Swindler. 1982. Growth of the mandible in adolescent pigtailed macaques *(Macaca nemestrina)*. *Human Biol.* 54: 31–44.

Smith, R. J., and M. M. Minium. 1983. Primate models in surgical orthodontics. *Am. J. Orthod.* 83: 235–244.

Solow, B. 1966. *The Pattern of Craniofacial Associations. Acta Odont. Scand.* 24 (suppl. 46).

Tanner, J. M. 1962. *Growth at Adolescence*. 2d ed. Oxford: Blackwell.

Turpin, D. L. 1968. Growth and remodeling of the mandible in the *Macaca mulatta* monkey. *Am J. Orthod.* 54: 251–271.

van't Hof, M. A., M. J. Roede, and C. J. Kowalski. 1977. A mixed longitudinal data analysis model. *Human Biol.* 49: 165–179.

Watts, E. 1980. Patterns of postnatal growth in the long bones of rhesus monkeys. *Am. J. Phys. Anthropol.* 52: 291 (abstract).

Watts, E. S., and J. A. Gaven. 1982. Postnatal growth of nonhuman primates: The problem of the adolescent spurt. *Human Biol.* 54: 53–70.

Weijs, W. A., and T. K. van der Wielen-Drent. 1983. The relationship between sarcomere length and activation pattern in the rabbit masseter muscle. *Arch. Oral Biol.* 28: 307–315.

Zerbe, G. O. 1979a. Randomization analysis of the completely randomized design extended to growth and response curves. *J. Amer. Stat. Assn.* 74: 215–221.

Zerbe, G. O. 1979b. Randomization analysis of randomized blocks extended to growth and response curves. *Communication in Statistics, Theory and Methods* A8: 191–205.

Zerbe, G. O. 1979c. A new nonparametric technique for constructing percentiles and normal ranges for growth curves determined from longitudinal data. *Growth* 43: 263–272.

Zerbe, G. O., and S. H. Walker. 1977. A randomization test for comparison of groups of growth curves with different polynomial design matrices. *Biometrics* 33: 653–657.

AUTHOR INDEX

AUTHOR INDEX

SUBJECT INDEX